MAN IN A COLD ENVIRONMENT

MONOGRAPHS OF THE PHYSIOLOGICAL SOCIETY

Number 2

Editors: L. E. Bayliss, W. Feldberg, A. L. Hodgkin

MAN IN A
COLD ENVIRONMENT

PHYSIOLOGICAL
AND PATHOLOGICAL EFFECTS OF EXPOSURE
TO LOW TEMPERATURES

by

ALAN C. BURTON, Ph.D.

Professor of Biophysics, University of Western Ontario

and

OTTO G. EDHOLM, M.B., B.S.

Head of Division of Human Physiology, National Institute for Medical Research,
Medical Research Council, England
(Formerly Professor of Physiology, University of Western Ontario)

(Facsimile of the 1955 Edition)

HAFNER PUBLISHING COMPANY

NEW YORK   LONDON

1969

BURTON: MAN IN THE COLD ENVIRONMENT

Copyright © 1955 Edward Arnold (Publishers) Ltd.
Reprinted by Arrangement 1969

Printed and Published by
HAFNER PUBLISHING COMPANY, INC.
31 East 10th Street
New York, N.Y. 10003

Library of Congress Catalog Card Number: 79-100389

Printed in the U.S.A.

CONTENTS

LIST OF ILLUSTRATIONS

List of Illustrations

List of Illustrations

INTRODUCTION

The need for yet another book on the effects of low temperatures may reasonably be questioned, so a brief account of how this book came to be written appears to be desirable. In 1948, the Defence Research Board of Canada organized a conference at Toronto on the effects of cold on man, and in the discussion at that meeting, attended by research workers from the United States, Canada and Great Britain, there was widespread agreement that much of the wartime work, buried as it was in special military reports, would be lost, if someone who had been connected with the work did not publish it in a more available form.

Subsequently the Arctic Panel of the Defence Research Board raised the question again, and finally the authors contracted to prepare a book reviewing the wartime work, and bringing the subject as far as possible up to date.

The recent publication of two books on the same general subject, *The Physiology of Heat Regulation* edited by Newburgh, and *Temperature and Human Life* by Winslow and Herrington has made the task of writing this book much easier. Detailed description and explanation of technical points is adequately given in the first of these. Our hope is that there may be a place, in addition, for a more unified presentation of a consistent scheme of evaluating the problems of man in the cold than is possible to a large group of contributors. Also some aspects of the problem, particularly those of tolerance when heat balance is not possible, of the state of hypothermia, general and local, of the pathology of cold injury, and resuscitation from cold have hardly been discussed in any book. The new subject of 'acclimatization to cold' has been included because it is of such present interest, though as yet we have little data for man.

The book falls naturally into several divisions. In the first the physical and physiological problems involved in maintenance of a thermal steady state are discussed, with a scheme for assessing the thermal demand of the environment. In the second section the ways in which animals have met the problem are described, first in general terms, then in some detail as to physiological mechanisms in man. A section follows describing the consequences that result if heat balance is not maintained, leading eventually to pathology. Finally a chapter has been written suggesting the lines of future research that are indicated by our evident areas of ignorance.

We have tried to emphasize general principles rather than to give any detailed treatment of special cases, particularly in the discussion on clothing. There is already an extensive literature on this subject, mainly in the form of reports to Governmental committees, but there would be little point in citing them in a book of this kind. Again there are several ways of expressing the principles of heat exchange and of calculating the thermal demand of the environment. Our method is not necessarily the best, but we have emphasized it, in the conviction that it is worth sacrificing some broadness of view to secure consistency and simplicity of presentation.

In the course of preparation of the book it became obvious that some of the original aims would have to be modified. A very considerable number of wartime reports have been read, but the majority are not cited, as the physiological content of importance was generally found to be small. Fortunately, with a revival of interest in the subject, many of the wartime workers have returned to the field, so ensuring valuable continuity.

The literature on the various effects of cold is so extensive, it has been found impossible to cite every author. An attempt has been made to mention all the more important recent papers, but as all those who have ever had to prepare a review will know, as soon as the available literature has been read and a critical summary prepared, new publications will have appeared which modify previous work. We have therefore decided to make December 1951 the deadline. Some material published after that date has been included, but in most cases such material was available to us earlier in the form of a service report or by personal contact with the workers concerned. There has been no attempt to construct a complete bibliography as this has recently been prepared by the Office of Naval Research, and this bibliography has been of the greatest assistance to us. The late Dr. Geoghegan, working on behalf of the Royal Naval Personnel Research Committee had also been preparing a bibliography, and very kindly gave us permission to use his material. We have included in the text, data and discussion which have not been previously published. This was felt to be justified, as fragmentary information obtained in the course of *ad hoc* wartime research might be of some value, even though it would not obtain publication in the usual way.

However, the most important way by which we have obtained

information of the current research on cold has been by direct personal contact with workers themselves. We have had the remarkable privilege of being able to discuss and watch work in progress in laboratories throughout Canada, the United States and Great Britain. Personal discussion has also taken place with those working in Scandinavia, Germany, Yugoslavia, etc. This has been due to the arrangement made on our behalf by the Defence Research Board of Canada, and the co-operation of the heads of service laboratories in the United States and Great Britain.

The help we have received is therefore very considerable, and we would like to take this opportunity to thank our colleagues, who have been so generous with information and advice. Inevitably we feel we have not been able to do justice to the material at our disposal, and we hope that our colleagues will realize that this book is not really meant for those already working in the field. Many papers which should have been cited have been omitted and we are only too well aware of the deficiencies. 'We offer you the repast: Now choose for yourself what you will eat.' (Dante, *Del Paradiso*, Book 10, line 25.)

Throughout the book temperatures have been expressed in degrees Centigrade with the Fahrenheit equivalent in brackets, except for calculations that are illustrative only. It is probably too much to hope that all workers will employ the Centigrade scale exclusively. After the meeting of the International Physiological Congress at Copenhagen, an informal gathering of some forty physiologists interested in climatic physiology was held. It was there agreed that the best policy at present would be to express temperature in degrees Centigrade with Fahrenheit in brackets. We strongly recommend this procedure to our colleagues.

We owe a special debt to Dr. Morley Whillans, head of the Defence Research Medical Laboratories, Toronto; he has encouraged us to continue with this book, and it is in large part owing to his stimulus that the book now appears.

We are greatly indebted for secretarial assistance to Miss D. Ainsworth, Mrs. D. Walls and Mrs. T. Lovegrove, and for help in preparing the diagrams to Miss K. Ronni and Mr. W. Austen.

ACKNOWLEDGEMENT TO THE DEFENCE RESEARCH BOARD

The Defence Research Board of Canada commissioned the preparation of the manuscript of this book, and also paid all the expenses in connexion with the visits made by the authors to research laboratories in Canada and the United States. The book could not have been prepared without this assistance, and the authors wish to acknowledge their gratitude to the Defence Research Board for the very generous aid provided.

Figures have been reproduced from many sources, and we wish to thank the following authors and journals for permission to do so:—

Dr. E. F. Adolph, Dr. L. Alexander, Professor J. Aschoff, Professor L. P. Dugal, Dr. E. S. Fetcher, Dr. F. E. J. Fry, Dr. R. T. Grant, Dr. J. S. Hart, Dr. L. H. Hegnauer, Dr. L. P. Herrington, Dr. R. E. Johnson, Dr. N. H. Mackworth, Sir Frank Markham, Dr. J. Mead, Dr. G. W. Molnar, Dr. L. H. Newburgh, Professor E. E. Pochin, Dr. E. Robillard, Dr. P. F. Scholander, Dr. J. C. Scott, Dr. E. A. Sellers, Dr. J. T. Shepherd, Dr. P. Siple, Dr. C. R. Spealman, Dr. R. F. Whelan, Dr. C. P. Yaglou.

The American Journal of the Medical Sciences, American Journal of Physiology, Biological Bulletin, Bulletin of the American Institute of Architects, Canadian Journal of Research, Clinical Science, Federation Proceedings, Heart, Irish Journal of Medical Science, Journal of the American Medical Association, Journal of Applied Physiology, Journal of Nutrition, Journal of Physiology, National Research Council, Canada, Oxford University Press, Pflügers Archives, Publications of the Ontario Fisheries Research Laboratories (University of Toronto Press), *Revue Canadienne de Biologie*, W. S. Saunders & Co., Philadelphia, *Science*, U.S. Department of Commerce, Washington (for the Combined Intelligence Objectives Sub-Committee).

CHAPTER 1

HOMEOTHERMY AND HISTORY

Life and Thermodynamics

To some, it is still a matter of debate whether the existence and the growth of living cells and organisms is a contradiction, though perhaps a merely local contradiction, of the generalized second law of thermodynamics. This law, which has been given in so many ways that the lay person is apt to be confused, states that the entropy of the universe is increasing to a maximum, and this implies that the universe is becoming more and more uniform and randomized. The energy of the universe is always tending towards 'degradation' into the ultimate random motion we call heat. While the law, which is statistical in nature, permits us to suppose that in small regions of the universe, small enough to be below the applicability of statistics, local decreases of entropy in areas of high organization and differentiation can arise by 'statistical fluctuation', it seems incredible that the differentiation and organization of even a single living cell could be explained on the basis of such statistical fluctuations, even when recourse is had to the principle of selection in evolution and the facts of genetic inheritance (though Schrödinger (1) seems to think this enough). Living cells and organisms are so fantastically improbable that there does not seem to have been time enough for them to arise by fortuitous 'experiment' under the second law of thermodynamics. Life is considered as 'disentropic' by many, such as Ubbelohde (2).

Yet while the existence and proliferation of life is, on this view, in defiance of the second law, which seems to apply to all of the inanimate world which we are able to observe, the mode of life of cells and organisms, once they exist, is definitely in accordance with that law.

A good analogy is the automobile, which, being the product of the mind and activity of living organisms, is also a wildly improbable assembly of highly selected and organized molecules, and another contradiction of the second law. Yet the operation of the automobile, the way in which it runs uphill as well as down, is strictly in accordance with the law of entropy. The chemical

energy of the fuel (petrol) is being degraded, a portion being transformed into mechanical energy but a great deal (over 70 per cent)
appearing as heat at once. Even the mechanical energy eventually
is degraded into heat in the friction of the moving parts and against
the air through which the car has moved. When the car is back in
the garage after the afternoon's drive, the end result is that all of
the low entropy energy of the petrol used has been transformed
into heat, and the entropy of the universe is correspondingly the
greater.

So it is with living animals. Their life depends upon the production of heat from the chemical energy of food. By their existence
they increase the 'flux of degradation' of energy in their locality.
Possibly this flux is greater than it would be if the world were all
inanimate, and living things somehow 'pay for' their special privi-
leges of emancipation from the second law, by their increased
contribution to the general trend.

As to the application of the first law of thermodynamics, the
conservation of energy principle, the long series of researches of
Lavoisier, Richet, Zunst, Lusk and a host of others culminated in
the work of Benedict and Atwater, DuBois and Murlin, to show
that, without a shadow of a doubt, all of the energy that appears
ultimately as heat is quantitatively accounted for by the chemical
energy of the food ingested. We will be concerned with both of
these laws of thermodynamics, in part with the first law in the
chapters on nutrition in the cold and in consideration of the heat
production of man, and a great deal with the second law, in consideration of the transfer of the degraded energy, the heat-loss,
to the universe about us.

The 'Excess Temperature' of Living Cells

One consequence of the second law of thermodynamics (and
this is the one emphasized in textbooks of thermodynamics since
originally the law was developed from considerations of heat
engines, long before the broad applications were appreciated) is
that heat flows only from regions of higher temperature to where
the temperature is lower. Thus the 'energy flux' of a living animal
cannot reach a steady state of exchange with its environment until
the temperature of the animal has risen above that of its surroundings. The temperature reached by the cells of the animal will be
determined by the turnover of energy of the cells and also by the
ease with which the heat so produced can diffuse to the surround-

ings, for the law of diffusion of heat states more than that it must flow from higher to lower temperatures (which is all the second law demands). The 'thermal gradient' down which the flow occurs is quantitatively related to the flow. The greater the heat produced by the energy turnover, the greater the excess of the temperature of the animal over that of its surroundings. We might depict the situation graphically as in Fig. 1*A*, where we have for simplicity assumed that the flow of heat is proportional to the excess temperature (this is approximately true). Thus the temperature of the tissues of an animal is determined by the activity or energy consumption of those tissues and by the way in which heat can flow down the thermal gradient, i.e. by the 'thermal insulation'. This is treated in more detail, and quantitatively, in succeeding chapters.

The Universal Effect of Temperature on Life

If the factors illustrated by Fig. 1*A* were all that must be considered, the study of the energy exchanges of animals and man would be fairly simple, though possibly rather dull. However, we cannot consider the rate of energy turnover as something determined by the animal, by its organization, enzymes and so on, and by that alone. The temperature of cells also directly affects their activity. The life of the organism depends upon the chemical reactions by which the entropic transformation of chemical energy into heat is proceeding. The rate of these chemical reactions, universally, is affected by their temperature.

The dependence of the rate of reactions upon the temperature is well expressed, for the great number of cases, animate and inanimate, that have been studied, by the well-known law of Arrhenius. This law states that the logarithm of the rate is proportional to the reciprocal of the Absolute Temperature. The law has no solid foundation of proof by thermodynamics, though kinetic theory has almost succeeded in giving it scientific respectability. The increase in velocity constant of a reaction, by this law, is not linear but rather logarithmic with temperature, so that as the temperature rises, the increase is accelerated. The temperature coefficient is often described by the Q_{10}, which gives the ratio of increase of rate with a rise of 10°C in temperature. For most metabolic processes the Q_{10} is between 2·0 and 3·0, i.e. the heat production of living cells will increase two to three times if their temperature is raised by 10°C. We may therefore depict the second relation between the energy flow of organisms and their temperature by a second graph, Fig. 1*B*. In this case we must use a curve rather than a straight line.

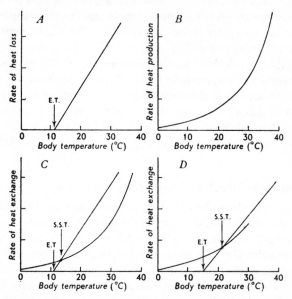

FIG. 1. Thermal exchanges in the poikilotherm. (*A*) Relation between heat loss and temperature of tissues. E.T. Environmental Temperature, taken in this case as 11°C. (*B*) Relation between heat production and temperature of tissues according to the Arrhenius relation. (*C*) Combination of *A* and *B*. S.S.T. Steady State Temperature, given by the intersection of the lines. (*D*) Increase of Steady State Temperature (S.S.T.) when the Environmental Temperature (E.T.) is increased from 11°C to 15°C.

Both of these relations must hold simultaneously, i.e., the temperature of the tissues is determined by the physical relation between excess temperature and heat loss, and also by the chemicophysical relation between temperature and heat production. Thus for the steady state we must combine the two graphs (Fig. 1*C*). The only possible steady state of the cells in these circumstances of environmental temperature is represented by the intersection of the two lines on the combined graph. The operation of these two laws, that of diffusion of heat and that of Arrhenius, impose a restriction on the freedom of the operation of the living animal. An animal cannot, according to these considerations alone, increase its activity without a change in the temperature of the tissues. Again, the temperature of its tissues, and therefore the biological activity of the animal, will depend upon the environmental temperature, since there is a new 'heat loss line' for each environmental temperature and a new point of intersection of the two lines (Fig. 1 *D*). Thermodynamics has therefore imposed upon the living organism a severe restriction of its freedom of action. The study of this restriction and how it has been modified by special mechanisms in some animals (the homeotherms) is the subject of this book. Other studies, such as the consequences of the restriction that are involved when a cell or whole organism grows, or the stability of the state represented by the intersection on the graphs (3, 4), are beyond the scope of this book.

The 'poikilotherm', not possessing the means to escape from these 'fetters of thermodynamics', is almost the slave of the environmental temperature. All that one has to do to reduce the dangerous alligator to a helpless state for physiological experimentation is to leave him in the ice-box for a few hours. We are told that the world was at one time populated by poikilothermic animals of fantastic size and strength, and presumably speed. Yet their life must have been at the mercy of the climate. On a cold day the battle of existence slowed to relative inactivity. Before the development of the special group of animals called homeotherms, possessing the power, within limits, to be emancipated from thermal slavery, the temperature must have played a much less important rôle in the story of evolution. The outcome would be little affected when both pursuer and pursued were slowed by a fall of temperature. But the puniest homeotherm, like the little shrew, could survive against its poikilothermic enemies, when it was able still to be active in the cold. Perhaps the dependence of the large poikilotherms upon the temperature was the chief factor in the extinction of most of their species, once they had to compete with others less powerful but enjoying thermodynamic freedom.

The Advantages of a Constant Brain Temperature

We have considered the advantages possessed by the homeotherm in terms of independence, within limits, of the tissue temperature from the environmental temperature. Such independence would not necessarily imply constancy of the deep body temperature. In the homeotherms we find that the deep body temperature, which is approximately the same for the 'core' of the body (e.g., deep rectum, viscera, liver, brain) is regulated to constancy within a remarkably narrow range of temperature; in the human about 36·4 to 37·5°C (97·5 to 99·5°F) with a diurnal rhythm.

It is of interest to inquire why such a constant temperature is desirable. A very reasonable hypothesis is that the greater the complexity of the integration of the organism, the greater is the need for constancy of temperature for efficient functioning. All chemical, and physical, reactions change their rate with change of temperature, but the degree of acceleration with rising temperature is different for different reactions. Extension of the theory of Arrhenius by the physical chemists predicts that the dependence of the velocity constant of a reaction upon temperature will depend upon the 'activation energy' (a measure of the energy required for the initiation of the reaction rather than the energy liberated when it proceeds). Activation energies in biological reactions vary widely. Thus when we have a complicated process involving the co-ordination of many individual reactions, a rise of temperature will not only speed the over-all rate but will alter the relative rates of the various reactions involved. If the brain were like a clock mechanism, we might imagine that a change of temperature would result in throwing the various reactions out of step with each other. The analogy is false, for the law of mass action applied to a complex of successive or interlocking reactions shows that there is a type of automatic stability by which the reaction rates are changed to keep 'in mesh' whenever one of them is altered. However, the relative concentration of all the reactants will be markedly different when the new steady state, at a different temperature, is reached, and thus side reactions, connected with these reactants, will change their relative importance. Change of temperature therefore produces not only a change of over-all rate of metabolic and other biological processes, it also changes the qualitative character of these processes. It is easy to see how the human brain, where we think the integrative co-ordination of many processes is most developed, will be greatly disturbed by change of its temperature, while simpler, more unitary processes, as in the brain of lower forms or the peripheral tissues of man, will be less affected. Experience confirms that only a slight change in brain temperature, as in fever (even in non-toxic, artificial fever) or hypothermia, does result in profound confusion in mental processes.

Confirmation of this idea that it is complexity of organization that makes homeothermy necessary to an animal is the interesting fact that very young animals, in which the full co-ordination and integration of the nervous system is not yet developed, show an astonishing tolerance to changes of their body temperature (5) together with a lack of temperature regulation. This tolerance is altogether greater than in the adult of the same species. Again when the cause of death in hypothermia in adult animals is investigated (6), it is apparent that this, which is usually cardiac in nature, is a failure of the co-ordination and integration of the organism rather than of any one of its parts. It would not be fair to conclude that increased susceptibility to thermal death is part of the price of homeothermy. Rather the susceptibility is a consequence of the increased complexity of organization of the adult higher animal, which also makes homeothermy a necessity for efficient functioning.

The Means of Thermal Emancipation—Physical and Chemical Regulation

Homeothermy is one example, and a most important one, of the general emancipation of higher organisms from the effects of changes in their environment. In all cases the means is by the provision of the constant 'internal environment' of Claude Bernard (7); see also discussion by J. Barcroft (8).

The means by which temperature regulation is achieved will be discussed in some detail in later chapters. The mechanisms are of two distinct classes, physiological, depending on changes in the animal body which are in general purely involuntary and reflex, and psycho-physiological or behavioural, involving conscious adaptations by the animal. The most important of the latter in the case of man is the provision of a private climate, by the use of clothing and shelter. Thus the range of environmental temperature over which homeothermy is possible has been greatly extended. Since we have used the graphical method to illustrate the restrictions imposed by the laws of thermodynamics and of chemical kinetics, it is of interest at this point to see how the methods of emancipation from these restrictions may also be illustrated on the same graphical scheme. The purely physiological methods only will be illustrated (though that of the use of clothing obviously could be similarly depicted). Let us suppose that for the animal the desirable range of deep body temperature (say brain temperature) was as in Fig. 2A, shown as 40°C to 37°C (104°F to 98·6°F). These are not of course the lethal limits, merely those at which the animal will function best. If there were no special regulatory apparatus,

B

the diagram would be as in Fig. 2A. The heat loss lines for these limits for the intersections with the heat production curves would define a very narrow range of desirable environmental temperature. If the environmental temperature was outside these limits, the body temperature would be outside its specified limits. Actually the range of desirable environmental temperature would be less in degrees than that postulated for the body temperature.

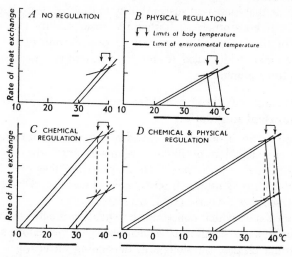

FIG. 2. Thermal exchanges in the homeotherm. The vertical scale for heat is in arbitrary units. (A) Showing that if the body temperature (S.S.T.) must be kept within narrow limits, the range of permissible environmental temperature is very narrow. (B) Increase in the range of permissible environmental temperature (E.T.) produced by physical regulation, by which the slope of the heat loss line can be changed. (C) Increase in range of environmental temperature by chemical regulation by which heat production can be increased (up to 3 times) for the same tissue temperature. (D) Combination of physical and chemical regulation. The small segments of curve shown are the relevant parts of the Arrhenius curves of heat production vs. body temperature.

The homeotherms possess what Rubner (9) called a 'physical regulation' of body temperature. The effective conductivity, or insulation, of the animal can be altered physiologically, so that for the same excess temperature of the body above the environment, the heat loss can be altered. This is by the two means, control of the circulation to the surface of the body, and control of the evaporation of moisture from the body. As a result we have (Fig. 2B) not one heat loss line, but two lines of quite different slope. The range of slope of the heat loss lines in Fig. 2B is roughly that of physical regulation in man, but not accurately so, nor does it take account of several complications. As a result there is a greatly widened range of environmental temperature within which the intersection of heat loss and heat production lines can still be within the limits of desirable body temperature.

'Chemical regulation' is a means by which the animal can greatly increase its heat production, even though the temperature of the tissues is the same. In animals and man the range, for continuous operation, appears to be to about three or four times the lowest heat production. Thus instead of one heat production curve on the chart (illustrating the Arrhenius equation) we have a series of curves at different levels (Fig. 2C). Chemical regulation is effected through the operation of the neuromuscular apparatus of the animal and also by hormones. Again, the range of desirable environmental temperature is greatly extended.

When both physical regulation, i.e. control of heat loss, and chemical regulation, i.e. control of heat production, are available, the range is still further extended. By the use of behavioural biological adaptations, posture, clothing, shelter, and air conditioning man has extended the range still more.

The Cost of Homeothermy, and the True Hibernants

The preceding discussion of temperature regulation shows that in order that an animal shall maintain its body temperature in cold environments, its heat production must be continuously at a high level. This imposes a restraint upon the freedom of the homeothermic animal, which partly removes the advantages of emancipation from the thermal environment. A sustained high level of heat production, by the law of conservation of energy, demands that the animal consume more food and thus more time and energy must be spent in obtaining it. In this sense there is a considerable cost to pay for homeothermy. In the case of man this cost, which may become excessive in very cold climates, has been partly evaded by the biological adaptations of the more advanced level, concerned with behaviour. These are the use of clothing, shelter, auxiliary heat (fire, etc.), and in modern times, air conditioning. In animals not possessing the means of evading this cost of homeothermy in the cold, whose evolution is still in the purely physiological stage, rather than psycho-physiological, the advantages of homeothermy will disappear with the cold season. Indeed many of the arctic homeotherms may lead a marginal existence in the winter months, when their increased need for calories in food is coincident with a decrease in the availability of that food.

Some animals, capable of homeothermy in warm or temperate environments, have met this difficulty by abandoning homeothermy altogether in the cold. These are the 'true hibernants', of which the best-known examples are the European marmot, the American ground-hog, and some ground squirrels. The bear, so often referred to in this connexion, is definitely not a true hibernant, but remains warm-blooded in his winter sleep. In their winter sleep, the true hibernants are poikilothermic, with their deep body temperature only a degree or two above the environmental temperature, and their metabolism profoundly depressed. Reference should be made to Johnson's review on the subject (10). One feature of true hibernation should be noted, which distinguishes it completely from the dormant state of poikilothermic animals in the cold, often called hibernation, and the similar state of homeothermic animals if rendered hypothermic. Though most reflex activity and irritability is profoundly depressed, there remains in the sleeping hibernant a protective reflex of great interest. If the environmental temperature falls below the levels which would produce freezing of the tissues, the animal will wake and in a few hours will pass through a remarkable process of rewarming back to the warm-blooded state. Indeed this operation of the protective reflex is probably the explanation of the emergence of the ground-hog from his burrow on or about 'ground-hog day' (February 2nd) to 'see his shadow' and influence the coming of spring. For the temperature in the ground at the depth of his burrow (4–6 ft.) lags considerably behind that at the surface, and is likely to reach its minimum at that time (11).

A second difference to be noted from the non-hibernant homeotherm is that the temperature regulation of the hibernant, when in the warm blood state, is of very much inferior accuracy and sensitivity (12). The deep body temperature fluctuates over quite a wide range, depending on the activity of the animal as well as the environmental temperature. The level of body temperature regulation is also in many cases significantly lower than that of the non-hibernating animals, which as a group show a remarkably small interspecies range. In the case of the bat (13) it is doubtful whether we should classify this animal as being homeothermic at all, since in the periods of sleep the body temperature drops to a few degrees above the environment and in the periods of wakefulness and flight rises towards 40°C (104°F). The bat seems really to be a poikilotherm capable of unusually great metabolic activity. In view of these fundamental biological differences between the hibernants and the non-hibernating homeotherms, it does not seem likely that we can learn much from the study of hibernation that will suggest application to the problem of homeothermy in the cold. Fundamental research however might be directed to one feature of the biochemistry of hibernation. The heart of a marmot, if removed when the animal is in the warm blooded state, appears just as susceptible to arrest when cooled as that of other homeotherms (say the dog). Yet it is reported that the excised heart of the marmot that is hibernating can continue to beat like that of a frog. This suggests a fundamental biochemical difference in the tissues (14 and 15 a, b, c, d). The level of magnesium, or of sugar, of insulin or other hormones, suggest themselves as relevant, but there is no conclusive evidence of their rôle.

Though the human has no mechanism by which his homeothermy can be abandoned, so that he may live successfully as a poikilotherm when the cost of homeothermy becomes excessive, he makes instinctive adjustments in this direction. After the first World War, an extensive survey was made by a team of experts to assess the deleterious effects of under-nutrition on European populations (16). These were found to be absent to a remarkable degree. It was suggested that the habit of people, in cold winters and lacking in food supply, of spending all day in bed with heavy covers, might explain how they had evaded the consequences that were expected.

The Level of the Regulated Temperature—Why 37°C (98·6°F)?

A survey of the deep body temperature of various species of homeotherms which has been measured in a great number (17), shows that the whole range is from 36°C (96·8°F) for the elephant, to 41°C (105·8°F) for the birds. It has been a matter of general speculation why we should not find homeothermic animals that regulated their 'core temperature' at lower or at higher temperatures, say at 33°C (91·4°F), or 45°C (113°F). Few suggestions indeed have ever been advanced.

It may be pointed out that the optimum temperature for the activity of many enzymes of biological importance is about 37°C (98·6°F) though in fact many exist in the body with optima at higher or lower temperatures than this. The suggestion is that biological adaptation has been towards such an optimal temperature for enzymes. This ignores the fact that a very large part of the body tissues, where enzymes function, are usually at considerably lower temperatures than 37°C (98·6°F) (see page 14). Also we know that many slight variants of enzyme structure are possible giving new enzymes of similar function but having different optimal temperatures. In microbiology, remarkable adaptations of this kind are familiar. It would seem much more likely that the enzymes of the body have been adapted to their internal environment, rather than that they have determined the level of body temperature.

A theory of some plausibility, but certainly of only speculative importance, is possible. Since, as we have seen, homeothermy demands that adjustments be made in the heat loss and/or the heat production when the body temperature changes, to restore it to the normal level, it might be argued that the adjustment will be made most easily if the thermal steady state should possess a kind of automatic stability. We have seen (Fig. 1) that a rise of body temperature will increase the heat loss to the environment (supposing that temperature to be constant) and will also increase the heat production. It may be that there is an advantage, to regulation, if the two were increased equally by the rise of body temperature so that they remained in balance, and that then the body temperature could be restored most easily by compensatory factors in the animal. Whether this be completely logical or not, let us see what level of body temperature will give this stability of the heat balance. Graphically the condition is that the two lines on Fig. 1*C* are tangent to each other.

Taking the Q_{10} of metabolism as 2·5, a rise of 1°C would increase the heat production by $(2·5)^{1/10}$ or by about 9·6 per cent (a Q_{10} of 2·3 would give an increase of about 8·6 per cent for 1°C). To see what the rise of 1°C will do to the heat loss, we must assume some universal average temperature for the surface of the earth, or rather for that part of it in which the homeothermic animals developed. From a map giving the mean average temperature over the globe, an average can be calculated for the land masses. The mean from this would appear to be about 16°C (60°F). If, however, we exclude those regions where the mean temperature in the coldest months (say January for the northern hemisphere and July for the southern) is below freezing, the mean is about 23°C (73°F). Certainly the areas where civilization is known to have arisen and where (let us assume for the sake of the speculation) the development of homeothermic animals took place, lie in a large belt, including North Africa, Persia, Northern India, Southern China, Peru, Chile, Argentine, Southern Brazil, Southern Central Africa and Central Australia. This belt is where the mean annual temperature is between 21° and 26°C (70° and 80°F). Of course in our real ignorance as to the climate of the past and the geographical origin of species, any temperature between such limits would be as good as another. It might be suggested that 25°C would be a convenient figure to use. If x be the body temperature, then the excess temperature will be $(x - 25°)$ and this will increase to $(x + 1 - 25°)$ or $(x - 24°)$ for a degree rise of body temperature. Since the law of heat loss is that it is approximately proportional to the excess temperature, the heat loss will increase in the ratio $\dfrac{x - 24}{x - 25}$. Then for optimum stability of heat balance, we should have

$$\frac{x - 24}{x - 25} = 1 \cdot 096 \ (Q_{10} = 2 \cdot 5), \text{ or } 1 \cdot 086 \ (Q_{10} = 2 \cdot 3).$$

Solving this equation for x, we find: $x = 35°$ or $38°C$ ($95°$ or $100 \cdot 4°F$). The proof would certainly not stand up to the many criticisms that will be at once advanced. The only feature of value perhaps is the suggestion that the level of body temperature adopted by the homeotherms has something to do with the *stability* of temperature regulation.

It may also be suggested that the range of body temperatures actually found is a compromise between two disadvantageous temperature ranges. A body temperature regulated at a low temperature would mean that the excess temperature would be small and changes of environmental temperature would call for relatively greater proportional changes in the physiological factors. On the other hand there is a good deal of evidence of a lethal temperature at about 43°C (110°F) (18). At such fever temperatures it looks, from experience with artificial hyperthermia, where no bacterial toxin is involved, as if the body temperature is liable to rise spontaneously still higher. A further rise might increase the heat production, by the Arrhenius law, more than the heat loss and so perpetuate itself. (The reader with the mathematical mind will recognize that this amounts to the basis of the calculation already given.) The range from 36° to 41°C (105·8°F) would represent a useful compromise.

Homeothermy is Both Partial and Limited

It is unfortunate that textbooks of Physiology and of Biochemistry, when discussing the properties of isolated mammalian tissue, refer these properties to a standard temperature of 37° or 37·5°C (98·6° or 99·5°F), as though this were the physiological temperature of all tissues. Yet it has been known, long before Bazett and McGlone (19) measured them accurately, that there are gradients of temperature from the deep tissues to the skin, where, even in environments of comfortable temperature, the average temperature is considerably less than 37°C (98·6°F), about 33°C (91·4°F) in a man at rest in a room at 21°C (70°F). The surface temperature of the extremities is normally much less than this average surface temperature. Measurements with needle thermocouples thrust into the tissues show that the deep 'core' temperature is only reached at a depth of an inch or more. Calculation shows that more than 50 per cent of the tissues of man are within one inch of his body surface (20). As good a case could be made for studies of excised peripheral tissue at 33°C (91·4°F) as at 37°C (98·6°F). The temperature of choice should of course be related to the temperature of that tissue in life, which depends on its location in the body. An illustration of this neglect of the fact that homeothermy is partial is that the effects of carbon dioxide tension on the dissociation curve of blood and its exchanges of gases with the tissues are always taught in some detail, while the equally great or greater shift of the curve according to the temperature of the blood, as it reaches peripheral tissues, is seldom even mentioned.

The temperature of the deep tissues, the brain, heart and abdominal viscera such as the liver is not exactly the same; for example that of the liver is usually up to a degree higher than the deep rectal temperature. DuBois (21) has discussed the many different temperatures of the human body and its parts. These are close enough to uniformity and constancy to justify the simplifying concept of a central deep 'core' of the body of uniform, regulated, temperature, surrounded by a 'shell' of cooler peripheral tissues, whose temperature moreover is dependent on that of the environment as well as on physiological factors. Indeed, the homeothermy of the 'core' is accomplished, in great measure, by the adjustment of the temperatures in the 'shell'. Homeothermy of the whole animal, up to the surface itself, would be much more difficult to achieve, since the 'physical regulation' would be eliminated (except for sweating) and only 'chemical regulation' would remain. The variable temperature of the peripheral tissues leads to some disadvantages in the functioning of animals (athletes are not at their best until they have 'warmed up') but the effect is not as serious as the confusion which would result in the more elaborately organized parts of the core of the animal. We shall see that peripheral tissues do display special protective mechanisms, operating against damage by extreme hypothermia and freezing.

The concept of 'core' and 'shell' tissues is a crude one, though it is very useful. As our knowledge advances of the details of physiological adaptation to the cold, particularly of those changes which take place in 'acclimatization', we will probably have to recognize the importance of shifts in the relative volumes of the 'shell' and the 'core', and must abandon thinking of them as distinct and fixed subdivisions. A shift in the depth to which shell-temperatures extend would alter the 'stored heat' of the tissues which is available to buffer the departure from the thermal steady state which the animal in natural surroundings must continually be experiencing. Thus if the 'core' withdrew deeper within the body, the animal could tolerate longer exposure to conditions where there was a negative heat balance, and thus could acquire a greater heat debt, before making it up by change of environment or increased heat production* (22).

That homeothermy is also limited, in the range of environmental temperatures in which it can be achieved, has already been discussed (Fig. 2). In man, without the supra-physiological adaptations discussed in the next paragraphs, it is doubtful that homeothermy would be possible much below freezing temperatures, though the Tierra del Fuegans (23) seem to have endured this environment successfully without clothing. On the other end of the range of environmental temperature the limit is reached very soon, though here the humidity with the temperature (as combined in the 'wet-bulb temperature') is of great importance. The consequences of the limitations of purely physiological homeothermic mechanisms are clearly traced in the history of civilization.

---

* This idea is due to a conversation with Dr. L. D. Carlson, of the Department of Physiology, University of Washington.

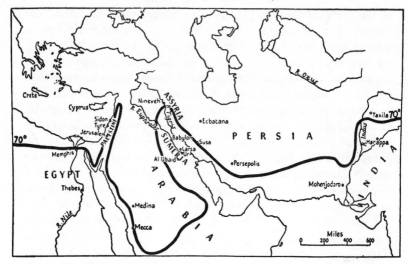

Fig. 3. The 70° isotherm and the sites of ancient civilization. For simplification, the therm is shown as a line, but it is rather a succession of areas. (From Markham, *Climate and the Energy of Nations*. Oxford University Press, 1944.)

Homeothermy and Civilization

Markham (24) has pointed out that the earliest civilizations of which we know all had their location close to the present 21°C (70°F) isotherm (Fig. 3). He investigates the evidence that this present isotherm is not very different from that which might have been drawn at the dawn of human civilization and concludes that changes in climate have occurred but not sufficiently to invalidate his argument. The apparent exception to his generalization is the early civilization in Peru where the mean temperature is considerably lower (perhaps 15·5°C (60°F)). These regions are highlands, however, where the radiation of the sun is unusually great, and this factor must be added as an effective increment to the temperature (see Chapter 7).

The spread of civilization from the 21°C (70°F) isotherm to the cool temperature regions adjoining, such as the northern shores of the Mediterranean Sea and northern China and India can be directly linked with the discovery by man of the use of fire to create a 'private environment'. The invention of the fireplace and the discovery of coal did more to open new regions for civilization than any other factor. Thus the further conquest of the environment by man has been dependent on the supra-physiological factors of acquired and transmitted skills in the use of clothing, shelter, heating of dwellings and eventually air conditioning. These have enormously increased the range of environments in which homeothermic man can live. It is interesting to compare the concentration of early civilization near the 21°C (70°F) isotherm with the estimated present distribution of the world's population with respect to mean environmental temperature. This can be done from figures given in an appendix in Markham's book, which gives the population and mean annual temperatures of the countries of the world. The results are:

Mean Annual Temperature	−1–4·5°C 30–40°F	4·5–10°C 40–50°F	10–15·5°C 50–60°F	15·5–21°C 60–70°F	21–26·5°C 70–80°F	26·5–32·5°C 80–90°F
Millions of People	40	378	394	431	460	61

The mean population temperature from these figures is now 16·5°C (61·1°F). If we excluded the very large item of the 352 millions of India with a mean annual temperature of 26°C (79°F), the mean would be only 14·4°C (57·8°F). We see therefore that there has not only been a spread of population to each side of the 21°C (70°F) isotherm, but the centre of population has shifted to a region nearly 5·0°C (9°F) lower than where civilization

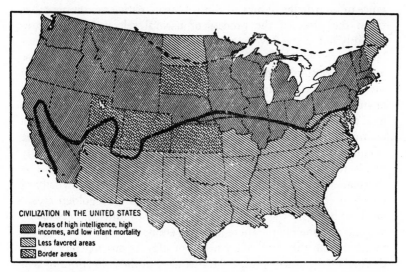

CIVILIZATION IN THE UNITED STATES
- Areas of high intelligence, high incomes, and low infant mortality
- Less favored areas
- Border areas

Fig. 4. Climate and civilization in the United States. North of the broken line the coldest month has a mean below −12°C (10°F). South of the heavy line the warmest month has a mean above 24°C (75°F). (From data of Huntington used by Markham and adapted.)

arose. Mere population figures are of course not good indices of 'civilization'. Markham devises various tests of what he calls the 'energy of nations', which include infant mortality rates, death rates, national incomes and share of world trade. He then shows that a remarkable correlation, which of course might still be accidental, exists between the areas of high national energy, so measured, and climatic conditions. Figs. 4 and 5 illustrate that this is apparent even within one country, and within one as small as England. Huntington (25) has also written extensively on this subject. It emerges from these studies of the 'energy of nations' that the most vigorous civilizations are those living in climates with mean annual temperatures from 6° to 16·5°C (43° to 62°F), the mean of the list is 10°C (50°F). Markham's leading countries are New Zealand, Australia, the Netherlands, Canada, United Kingdom, Denmark, the United States and Sweden. New Zealand heads every list. In none of these countries does the mean temperature for the hottest month exceed 21·5°C (71°F) for the populated areas, nor the coldest month fall below −2°C (28°F) except in Canada and the United States where heating and air conditioning systems are so well developed.

Thus in human history, the new tool of evolution, the inheritance of knowledge of how to control the climatic environment, has taken the place of the process of natural selection which presumably led to the development of homeothermy. We may speculate as to the reason for the shift of the maximum energy of peoples which is continuing now, to colder and colder environments, and for the decline of relative importance of civilizations in the ideal environments where they arose. The answer is possibly that given by Toynbee (26). Man to reach his full powers needs the spur of a continual 'stimulus' from his environment. This stimulus must be adequate to call forth his energies, but not so great that the majority of his time and activity is taken up with dealing with it. The thermal environment may supply such a stimulus, the intensity of which in the cool temperate regions is near 'the golden mean' of Toynbee. In the Arctic regions the stimulus is obviously so great that men living there have not enough energy left to devote to other than the elementary needs of life. As our mastery of the thermal environment grows, the adequacy of the cold stimulus will decrease, and we may expect the energy evoked in the cool temperature regions to decline, and foretell that the more vigorous peoples will move farther north. Gilfillan (27), noting the 'cold-

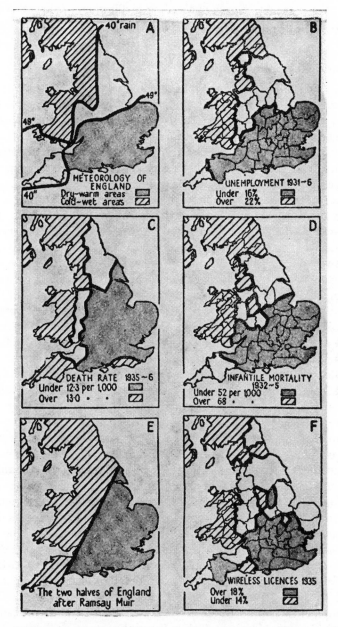

FIG. 5. Climate and energy in England. (From Markham, *Climate and the Energy of Nations*. Oxford University Press, 1944.)

ward' movement of each succeeding civilization, predicts that by
A.D. 2100, Montreal and Oslo will be the focal points of civilization.

Practical Applications

In view of the last paragraphs it would be easy to conclude that,
since the extension of man's conquest of cold by clothing, shelter,
and artificial heating has been so great compared to that accomplished by purely physiological mechanisms, the study of the
physiology of man in the cold is now of little practical importance. This conclusion would be foolish in view of general experience of applied physiology during the last war. In almost every
case of the general problem of enabling men to survive and
function efficiently in adverse environments, as at high altitude, the
answer to the problem proved to be in the field of engineering,
rather than in modifying human physiology. For example, the
answer to the problem of anoxia was not to attempt to increase the
human tolerance to anoxia, though methods of doing this were
suggested and were the subject of intensive research early in the
war, but to provide the man with oxygen under sufficient partial
pressure to ensure that there was no anoxia. Yet the solution was
not available until fundamental physiological research had established how much oxygen was required under different conditions,
under what pressures in different phases of the respiration and so
on. The engineering answer is not perfected without the fundamental knowledge of the underlying physiology. We may be
sure the same principle applies to research on man in a cold
climate. Conquest of this environment is possible only by the use
of clothing, shelter, and the use of auxiliary heat, but without
fundamental knowledge of physiological temperature regulation,
its mechanisms and its limitations, these devices will not be developed to their full possibilities. Again war-time experience
showed that ultimately success in living in adverse conditions de-
pended as much, if not more, upon the knowledge of how to
behave and use the protective equipment available, than upon that
equipment itself. An example was the observation of a marked
difference in the incidence of casualties due to trench-foot and
cold in one division in France and the division next to it. The
equipment was probably more highly developed and more readily
available in the division which suffered the greater casualties. The
difference was due to ignorance in the one case and knowledge in

the other of the 'tolerance time' of troops in those conditions, of how long men should be kept in the front line before relief, how often socks should be replaced and so on. Knowledge of the biological factors, including the behavioural and physiological, is paramount to successful achievement of adaptation.

REFERENCES

1. SCHRÖDINGER, E. 'What is life? The Physical Aspects of the Living Cell.' Macmillan, London, 1945.
2. UBBELOHDE, A. R. 'Time and Thermodynamics.' Oxford University Press, London, 1943.
3. BURTON, A. C. The Properties of the Steady State, Compared to those of Equilibrium as shown in Characteristic Biological Behaviour. *J. Cell & Comp. Physiol.* **14**, 327, 1939.
4. BURTON, A. C. The Application of the Theory of Heat Flow to the Study of Energy Metabolism. *J. Nutrition*, **7**, 481, 1934.
5. ADOLPH, E. F. Cardiac Responses to Hypothermia in Infant Mammals. *Am. J. Physiol.*, **163**, 695, 1950 (Abstract).
6. HEGNAUER, A. H. and PENROD, K. E. Observations on the Pathologic-Physiology in the Hypothermic Dog. Air Force Technical Report No. 5912, U.S.A.F. Air Materiel Command. Wright-Patterson, A.F. Base, Dayton, Ohio, 1950.
7. BERNARD, C. 'The Phenomena of Life.' Paris, 1878.
8. BARCROFT, J. Features in the Architecture of Physiological Function. New York, The Macmillan Company; Cambridge, Eng., The University Press, 1934.
9. RUBNER, Energiegesetze, 1902. See Lusk, G. 'The Elements of the Science of Nutrition', Chapter 5, Saunders, Philadelphia, 1928.
10. JOHNSON, C. E. Hibernation in Mammals. *Quart. Rev. Biol.*, **6**, 439, 1931.
11. SIMPSON, S. The Relation of External Temperature to Hibernation. *Proc. Soc. Exper. Biol. and Med.*, **10**, 180, 1913.
12. SIMPSON, S. Temperature Regulation in the Woodchuck. *Am. J. Physiol.*, **29**, 12, 1911.
13. BURBANK, R. C. and YOUNG, J. Z. Temperature Changes and Winter Sleep of Bats. *J. Physiol.*, **82**, 459, 1934.
14. LUSTIG, B., ERNST, T. and REUSS, E. Blood Composition in Summer and Winter of Felix Pomatia. *Biochem. Z.*, **290**, 95, 1937.
15. SOUMALAINEN, P. (a) Magnesium and Calcium Content of Hedgehog Serum during Hibernation. *Nature*, **141**, 1938.
    (b) Production of Artificial Hibernation. *Nature*, **142**, 1157, 1938.
    (c) Artificial Hibernation. *Nature*, **144**, 443, 1939.
    (d) Hibernation of the Hedgehog. 6. Serum Magnesium and Calcium. *Ann. Acad. Sci. Fenn. A.*, **53**, No. 7, 1939.
16. MURLIN, J. R. Personal Communication, 1938.
17. PEMBREY, M. S. 'Animal Heat.' Section of 'Textbook of Physiology.' E. A. Schafer. Vol. 1. Hodder and Stoughton, London, 1898.
18. DuBOIS, E. F. Why are Temperatures above 106°F Rare? *Am. J. Med. Sci.*, **217**, 361, 1949.
19. BAZETT, H. C. and McGLONE, B. Temperature Gradients in the Tissue of Man. *Am. J. Physiol.*, **82**, 415, 1927.

20. BURTON, A. C. The Average Temperature of the Tissues of the Body. *J. Nutrition*, **9**, 261, 1935.
21. DuBois, E. F. The Many Different Temperatures of the Human Body and its Parts. *Western J. of Surg.*, **59**, 476, 1951.
22. CARLSON, L. D. Personal Communication. 1950.
23. WULSIN, F. R. Adaptations to Climate among Non-European Peoples. Chapter 1 in 'Physiology of Heat Regulation and the Science of Clothing.' L. H. Newburgh, Saunders, Philadelphia, 1949.
24. MARKHAM, S. F. 'Climate and the Energy of Nations.' Oxford Univ. Press, London, 1944.
25. HUNTINGTON, E. 'Civilization and Climate.' Yale Univ. Press, New Haven, 1924.
26. TOYNBEE, A. J. 'Study of History.' Abridgement of Vols. 1–6 by D. C. Somervell. Oxford Univ. Press, London, 1946.
27. GILFILLAN, S. C. The Coldward Course of Progress. *Political Science Quarterly*, **35**, 393, 1920.

CHAPTER 2

THE PROBLEM OF THE HOMEOTHERM, THE HEAT-BALANCE AND PHYSICAL LAWS

1. The Heat-balance

The animal body is an 'engine' in the thermodynamic sense, in that it is continuously transforming energy from one form, the chemical energy of food, into other forms, namely mechanical work and heat. Inevitably the second law of thermodynamics applies, and a large part of the total energy which is transformed must appear in its final form as heat. If the temperature of the animal is to remain constant, the system must be in a thermal steady state in which a balance is maintained between heat production and heat loss to the environment. If heat production temporarily exceeds heat loss, the excess will be 'stored' as a rise of the average temperature of the tissues; if heat production is less than heat loss, there will be accumulation of a 'heat debt' in a fall of average body temperature.

It is unfortunate that the term 'heat-balance' has been generally adopted for this maintenance of equality between heat production and heat loss, for the word 'balance' suggests a state of equilibrium. The state, however, is not one of equilibrium but a 'steady state', in which there is a continual flux of energy through the system (in equilibrium the flux in one direction is balanced by an equal flux in the opposite direction). Both sides of the equation of the steady state, heat-production and heat-loss, are variable quantities, depending on both physiological and physical parameters. The heat production, however, varies much more with physiological than with physical environmental factors. Any dependence of the heat production on the physical environment is because the environmental temperature and humidity can change the temperature of the peripheral tissues, even if the 'core' temperature is kept constant, and so change the metabolism of those tissues. We must also remember that the environmental factors can alter the metabolism of the animal by reason of its psycho-physiological response to the environment, viz. voluntary physical exercise will be less in hot environments.

In contrast the heat loss depends more upon the physical factors of the environment than upon the physiological factors, though not overwhelmingly so. A great deal of confusion of thought existed before it was realized that the dependence of heat-loss upon the physiological factors on the one hand, and on the physical on the other, could easily be separated. The division is simply made, for with few exceptions the physiological variables operate entirely in controlling the flow of heat from the interior of the body to the skin, while the physical factors operate in the 'private climate' from the skin to the environment. The exceptions are that the physiological variables of posture and of bodily movement and also pilo-erection, can affect the heat loss from the skin. We therefore avoid a great deal of confusion when we separate the total gradient of temperature, that is, from 'core' temperature to the environmental temperature, into its two parts, namely the 'physiological gradient' (core temperature to skin temperature) and the 'physical gradient' (skin temperature to environmental temperature). The flow of heat in these two regions depends upon these two temperature gradients respectively. This is because of the fundamental laws of heat flow, discussed below.

2. Laws of Heat-flow

These laws have been so well described by Hardy (1) in his article in Newburgh's book, that there would be no point in repeating the arguments. The purpose here is to bring up, and if possible to clear away, some difficulties, and to substitute for the accurate but almost unusable detailed equations there given, a simple logical scheme by which heat exchanges can be treated, without too much loss of accuracy. The scheme is one which has been widely applied in practical research on human thermal engineering.

The fundamental law which has been generally employed in treating the problem of the animal heat-balance is *Newton's law of cooling*. This is an *empirical* law, resulting from laboratory experiments by Sir Isaac Newton, which states that the rate of cooling of a body which is hotter than its surroundings is proportional to its surface area and to the 'excess temperature', i.e. to the difference of temperature between the body and its surroundings. The rate of cooling is proportional to the heat loss (there is no heat production involved here). The constant of proportionality is the

constant thermal capacity of the body. Newton's law thus yields
the equation:—

$$H = C \times S\,(T - T_A). \quad \ldots\ldots\ldots\ldots\ldots\ 1$$

Where H is the rate of heat loss, say in kcal/hr, S is the surface
area, say in sq. m, T is the temperature of the body, say in °C,
and T_A is the temperature of the air. C is a factor, which is
constant in a given experiment, i.e., does *not* depend on the temperatures in the equation, though it does depend on other factors
such as the air movement and the shape of the body. C is called

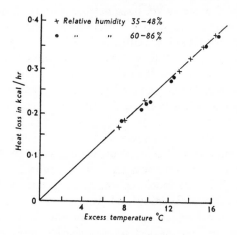

FIG. 6. Rate of heat loss of thermostated cylinder (interior kept at 37°C) in surroundings of different temperature and humidity. Note the independence of the
humidity, and the validity of Newton's law.

the 'cooling constant'. If there is no heat production in the body,
integration of Newton's law, in its form as a differential equation,
predicts that the temperature of the body will fall in an exponential
manner, finally to reach asymptotically the environmental temperature.

Newton's law is an approximation that holds very well over an
astonishingly great range of environmental conditions. How well
it does so is illustrated by Fig. 6, which gives the heat loss of a
heated, thermostated can of water in an air conditioned room held
at different levels of temperature and humidity (but constant air
movement). The introduction of this figure is convenient here
because it also illustrates a little known fact about Newton's law.
This is that the cooling constant C in air depends neither on

temperature *nor on the humidity* of the environment. The popular idea that it will alter with humidity is doubtless based on the knowledge that our sensations of cold are considerably affected by the humidity of the environment. 'Damp-cold' is universally thought to be 'colder' than 'dry-cold'. This is not to be denied, but there is a great difference between the observations of a clothed man, who moreover depends on evaporation of moisture from the skin for a considerable part of his heat-loss, and the 'naked', dry, inanimate can of this experiment. Nevertheless this difference between the sensations in damp-cold and dry-cold poses a real problem, of which we as yet have not the answer. Logic suggests that it must have to do with the properties of clothing, and that our observations would be different with the unclothed man. Unfortunately even for this we have no experimental data. This problem will be discussed later. Here the emphasis is that as far as loss of heat by radiation and convection are concerned, Newton's law does apply and has been verified by such experiments many times (2).

The form in which the equation of heat loss has been given is that used by engineers. Tables of the cooling constant C for pipes of different diameter and surfaces, with different air movements, are to be found in handbooks of engineering data (as that of the American Society of Heating and Ventilating Engineers). This form is most inconvenient, however, for the study of the heat losses of man, and we suspect for engineering problems also! A much more convenient form is

$$\frac{H}{S} = \frac{(T - T_A)}{I} \quad \ldots\ldots\ldots\ldots\ldots\ldots\ldots\ldots \ldots\ldots \quad 2$$

where we have simply replaced the cooling constant C by its reciprocal I. I is the 'thermal insulation', or if preferred by some, the 'thermal resistance'. It will be at once appreciated that 'thermal insulation' is the strict analogue of 'electrical resistance' which is defined by the analogous law of flow of electricity, Ohm's law. The reason why it is more convenient to use 'thermal insulation' is that more often we have to deal with thermal insulations 'in series' with each other, through which the same flow of heat passes in succession (as from the centre of the body, to the skin, through the clothing, and from the surface of the clothing to the air), than we have to deal with insulations in parallel. The same laws of summation for insulations in series, or in parallel, apply as in the case of electrical resistances. The total thermal insulation of several insulations in series is simply the arithmetic sum of the individual thermal insulations. For the case of insulations in parallel we have to invert, add reciprocals and invert once more. To avoid all this inversion and re-inversion we might just as well use the inverse of the cooling constant throughout. Sometimes, of course, we encounter problems in human heat exchanges where thermal resistances are in parallel, but these cases are less numerous.

One further step is required before we can apply this equation to the heat loss in the animal body down the physiological gradient, as well as down the physical gradient of temperature. Fourier and others developed the theory of flow of heat in solids, such as metals, and found there was a fundamental equation of the same form. The flow per unit area of cross-section was proportional to the difference of temperature, as in Equation 2. When the length of the thermal pathway is also unity, the constant C in the equation becomes the 'thermal conductivity' of the material. If the flow of heat from any point A to any point B was involved, the constant would be called the 'thermal conductance', per unit cross-sectional area, between A and B.

Now the flow of heat from the 'core' of the body to the skin is not purely by thermal conduction in the physical sense (this part would be present in tissue without bloodflow), but is mainly by 'convection', i.e. actual carriage of heat from one place to another by warm blood. This is why the transfer of heat in the tissues can be altered physiologically by a change in the peripheral blood flow. The constant of heat transfer for the tissues therefore depends on the blood flow mainly, and secondly on the thermal conductivity (in the physical sense) of the tissues. This will alter with the amount of fat, water, etc., in tissues. Also, it can be shown that even though heat generation in the tissues is a complication, the equation still holds (3). Therefore, this form of equation is universal, and we can state the whole problem in three equations.

(*a*) For the heat flow from the core of the body to the skin surface,

$$\frac{H_1}{S} = \frac{T_c - T_s}{I_T} \quad \dots\dots\dots\dots\dots\dots\dots\dots\dots 3$$

Where T_c is the core temperature, T_s the skin temperature, and I_T the 'insulation of the tissues'.

(*b*) For the heat flow from the skin to the external air, through the clothing, or fur of an animal,

$$\frac{H_2}{S} = \frac{T_s - T_{cl}}{I_{cl}} \qquad \dots\dots\dots\dots\dots\dots\dots 4$$

Where T_{cl} is the temperature of the outside surface of the clothing, and I_{cl} is the 'insulation of the clothing', or fur.

(c) For the heat flow from the clothing surface to the surrounding air,

$$\frac{H_3}{S} = \frac{T_{cl} - T_A}{I_A} \qquad \dots\dots\dots\dots\dots\dots\dots 5$$

Where T_A is the environmental temperature and I_A is the 'insulation of the air'.

These three insulations, of tissues, clothing, and air, are in series with each other. The quantity of heat that flows through the clothing is the same as that which flows into the air, i.e. $H_2 = H_3$. Unfortunately, the heat that flows through the tissues, H_1, is not equal to H_2 and H_3, if in the last two we are considering only the flow of 'non-evaporative' heat by radiation and convection. Heat is also lost from the skin by evaporation, and H_1, the heat that came up to the skin from the tissues, must include this evaporative heat as well as that denoted by H_2 and H_3. However, where there is no sweating, the evaporative heat loss from the skin is, as a good approximation, a constant fraction of the total heat loss of the skin. It has been shown (3) that on this assumption we can take H_1, as $1 \cdot 21$ times H_2 or H_3. Using the relation, we may eliminate the temperatures of the skin and of the clothing surface from the equations, and obtain for the total non-evaporative heat loss of the animal:

$$\frac{H}{S} = \frac{T_c - T_A}{1 \cdot 21\, I_T + I_{cl} + I_A} \qquad \dots\dots\dots\dots\dots 6$$

The insulation of the animal is then:

$$I = 1 \cdot 21\, I_T + I_{cl} + I_A \qquad \dots\dots\dots\dots\dots 7$$

Equation 6 shows us concisely the scope of the study of the non-evaporative heat loss of animals. Such study consists of an examination, in turn, of the three thermal insulations, that of the tissues, that of the clothing or fur, and that of the air, to determine their values and the factors upon which they depend.

The difficulty as to an assumption about the constancy of the evaporative loss as a fraction of the whole (in order to relate H_1 with H_2 and H_3) is avoided completely in work on protective clothing for man in a cold climate. Here we use equations 4 and 5 only, and measure the skin temperature, T_s, i.e. we consider the physical gradient of temperature only. We have then, without any assumption as to the evaporative loss, for the non-evaporative heat loss:

$$\frac{H}{S} = \frac{T_s - T_A}{I_{cl} + I_A} \quad \dots\dots\dots\dots\dots\dots\dots\dots\dots\dots 8$$

However, we need to take the full equation 6 when we consider all the means by which the heat loss of animals can be controlled.

3. The Evaporative Loss of Heat

Of the total heat loss of man, at least 20 per cent is usually by evaporation of moisture, both from the skin and from the respiratory tract. In basal, non-sweating conditions, the evaporative loss (insensible perspiration) is a remarkably constant fraction of the whole, from 23 to 27 per cent according to DuBois (4). The total moisture loss is normally about one-third from the respiratory tract, two-thirds from the skin surface. In hot conditions where the total heat loss has to be greatly increased, the chief means of such increase in man is by secretion of sweat and a consequent increase of evaporation from the skin, if the physical conditions are such as to allow more evaporation. In animals which lack sweat glands distributed over the skin, like the dog who pants, the increase is on the other hand by increased evaporation from the respiratory tract. Panting is a specialized mechanism for increasing the evaporation from the tongue without serious increase of ventilation of the lungs and consequent loss of CO_2.

The evaporation is driven by the difference of vapour pressure at the skin and vapour pressure in the environment, and may be described by an equation:

$$\frac{E}{S} = \frac{(W\mu)}{100} (P_S - P_A) \quad \dots\dots\dots\dots\dots\dots\dots\dots 9$$

where E is the rate of evaporative heat loss, say in kilocalories per hour, P_S is the vapour pressure at the skin (saturated at skin temperature) and P_A the vapour pressure in the air, say in mm. Hg. $\frac{W\mu}{100}$ is the evaporative cooling constant. The equation was put in this form by Gagge (5) for application to the human body. μ is the constant for a totally wet surface of the same shape as the body surface, and W is then the 'per cent wetted area' of the body. Gagge showed that if there was no sweating, but only 'insensible loss', the wetted area W fell to a minimum of about 10 per cent of the maximum which would be reached when the body surface was totally wet by unevaporated sweat. The physics of the derivation of this equation, as of the similar ones for loss of heat by radiation and by convection, is fully discussed by Hardy (*Physiology of Heat Regulation and the Science of Clothing*, Chapter 3).

For the evaporation from the respiratory tract but not 'from the lungs', since there is evidence (6) that practically all, if not all, the evaporation is from the upper respiratory tract, the equation is

$$E = V\,(Q_{Sat.}\,T_C - Q_A) \times 0.60 \dots\dots\dots\dots 10$$

where V is the ventilation of the respiratory tract, say in litres/hr, $Q_{Sat.}\,T_C$ is the quantity of moisture (g/litre) in air saturated at core temperature. Christie and Loomis (7) and McCutchan and Taylor (8) showed that expired air was saturated at a temperature very close to deep body temperature. Q_A is the quantity of moisture in the inspired environmental air. 0.60 kcal/g of water evaporated is the recognized factor for the latent heat of vaporization in this case (see discussion by Hardy). Now since saturated vapour obeys approximately the gas laws, the quantity of water in saturated vapour is roughly proportional to vapour pressure ($PV = Q.RT$) and so the difference of vapour pressures ($P_S - P_A$) once more can appear in this equation for the respiratory evaporative loss as well as in the equation for evaporation from the skin. Thus it is possible to have faith in the form of the equation of Gagge for the total evaporation loss, which is identical with equation 9.

However, there is a logical difficulty about Gagge's equation which has apparently not been noted. W is a *physiological* variable, the 'wetted area'. How then can it multiply the second term of the brackets, i.e. P_A which is apparently a parameter of the external environment alone? This would constitute a denial of the 'principle of contiguity', i.e. it would argue 'action at a distance'. Also, the equation suggests, on first inspection, that we are to take the effective vapour pressure at the skin as the saturated vapour pressure at skin temperature (Gagge implies this in establishing his equation experimentally). How can this be so when, for example, the wetted area is only 10 per cent? The equation is nevertheless a sound one. Mathematically it assumes a deceptive form as an algebraical simplification of the logical equation. We should consider the skin surface as a mosaic of small areas which were 'wet', above which the vapour pressure was the saturated vapour pressure at skin temperature, separated by 'dry' areas over which the vapour pressure was that of the environment. The scale of magnitude of these areas and of the mosaic may be very small indeed. Perhaps, in insensible loss, the moisture emerges between the edges of the epithelial cells, but this is of no consequence. At a very small distance from the skin the diffusing vapour will have spread out laterally so that a uniform average vapour pressure is attained, and the details of the wet-dry mosaic are lost. This average value will be the only logical 'vapour pressure of the skin'. If W per cent is the wetted area, we would predict that this average vapour pressure of the skin would be

$$P_S = \frac{W.P_{Sat.}\ T_S + (100 - W)\ P_A}{100} \quad \ldots\ldots\ldots 11$$

where $P_{Sat.}\ T_S$ is the saturated vapour pressure at skin temperature over the 'wet spots' and P_A is the vapour pressure over the rest of the skin. For example at a skin temperature of, say, 33°C, $P_{Sat.}\ T_S = 37\cdot6$ mm. Hg. If W is not the 10 per cent value found for non-sweating skin equation 11 gives us a value for the vapour pressure of the skin, in an environment of 20°C, 50 per cent relative humidity ($P_A = 8\cdot8$ mm. Hg) of $\dfrac{(10 \times 37\cdot6) + (90 \times 8\cdot8)}{100} = 11\cdot7$ mm. Hg, which is very different from the saturated value of $37\cdot6$ mm. Hg suggested by the misinterpretation of Gagge's equation. We may calculate an illustrative table of the values of the vapour pressure close to the skin with different values of the percentage wetted area and of the environmental vapour pressure, as below. The results of such calculations can also be shown by a graph (Fig. 7). This shows how the level of skin temperature, in the normal range of 30° to 36°C, does not greatly affect the results, and how when the wetted area is low, the external humidity greatly affects the relative humidity at the skin. In contrast, if the wetted area is large, i.e. there is sweating, the external humidity has much less effect. Over the range of external vapour pressures in cool surroundings saturated vapour pressure at 20°C (68°F) is only $17\cdot5$ mm. Hg. An empirical equation that fits the curves well for the normal skin temperatures is:

Relative humidity at the skin $\% = W + \cdot027\ (100 - W)\ P_A$. 12

TABLE SHOWING THE VAPOUR PRESSURE AT THE SKIN FOR VARIOUS WETTED AREAS, ENVIRONMENTAL VAPOUR PRESSURES, AND SKIN TEMPERATURES.

Vapour Press. of air mm. Hg.	30°C			32°C			34°C			36°C			Skin temp. °C W in %
	10%	50%	80%	10%	50%	80%	10%	50%	80%	10%	50%	80%	
0	3·2	15·9	25·2	3·6	17·8	28·3	4·0	19·9	31·8	4·4	22·2	35·5	mm. Hg.
	10	50	80	10	50	80	10	50	80	10	50	80	% RH.
5	7·7	18·4	26·2	8·1	20·3	29·3	8·5	22·4	32·8	8·9	24·7	36·5	mm. Hg.
	24	58	83	23	57	83	21	56	83	20	56	82	% RH.
10	12·2	20·9	27·2	12·6	22·8	30·3	13·0	24·9	33·8	13·4	27·2	37·5	mm. Hg.
	38	66	86	36	64	85	33	63	85	30	61	84	% RH.
15	16·7	23·4	28·2	17·1	25·3	31·3	17·5	27·4	34·8	17·9	29·7	38·5	mm. Hg.
	53	74	88	48	71	88	44	69	87	40	66	86	% RH.
20	21·2	25·9	29·2	21·6	27·8	32·3	22·0	29·9	35·8	22·4	32·2	39·5	mm. Hg.
	67	81	92	61	78	81	53	75	90	50	72	89	% RH.
25	27·7	28·4	30·2	26·1	30·3	33·3	26·5	32·4	36·8	26·9	34·7	40·5	mm. Hg.
	84	90	95	73	85	93	66	81	92	58	78	91	% RH.
30	30·2	30·9	31·2	30·6	32·8	34·3	31·0	34·9	37·8	31·4	37·2	41·5	mm. Hg.
	95	97	98	86	92	96	78	88	95	70	83	93	% HR.
35	34·7	33·4	32·2	35·1	35·3	35·3	35·5	37·4	38·8	35·9	39·7	42·5	mm. Hg.
	(109)	(106)	(101)	98	99	99	89	94	97	78	89	95	% RH.
40	39·2	35·9	33·2	39·6	37·8	36·3	40·0	39·9	39·8	36·4	42·2	43·5	mm. Hg.
	(123)	(113)	(104)	(111)	(106)	(102)	(102)	100	99	82	95	98	% RH.

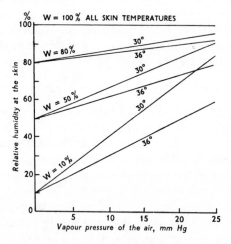

FIG. 7. Graph for estimating the relative humidity close to the skin from the percent wetted area (W), and the average skin temperature, at different vapour pressures of the environment.

where P_A is the environmental vapour pressure in mm. Hg. (It is to be noted that, rather unexpectedly, the relative humidity in the clothing can be greater than the relative humidity close to the skin (see chapter 4).)

However, the puzzle is soon resolved. The driving pressure-gradient from skin to environment is $P_S - P_A$

$$\text{i.e.} \quad \frac{(W P_{Sat} T_S + (100 - W) P_A) - P_A}{100} \quad \dots 13$$

which simplifies to $W/100 \,(P_{Sat} T_S - P_A)$ as in Gagge's equation. It seems worth raising the difficulty, only to remove it, because there has been the widespread misconception that the vapour-pressure on the skin is to be taken, in all circumstances, as the *saturated* vapour pressure at skin temperature. Actually, as shown above, it is usually much less than this. This view of the meaning of Gagge's equation is the same as that given by Mole (9).

Again it is much more convenient to change from an evaporative loss constant, such as $W\mu$, to its reciprocal, which will be a *resistance* to the transfer of water vapour—or simply 'vapour resistance', and write the equation

$$\frac{E}{S} = \frac{P_S - P_A}{R} \dots\dots\dots\dots\dots\dots\dots\dots\dots\dots\dots 14$$

where P_S is the average vapour pressure at the skin, given by equation 12, and P_A the vapour pressure of the environment, given by the relative humidity $\times$ saturated vapour pressure at air temperature. R is the total vapour resistance from the skin of the animal to the environment. Again, as with the total thermal insulation, we may consider the total vapour resistance R as made up of two resistances in series, the vapour resistance of the clothing or fur of the animal, R_{cl} (this will depend on the fibre, material design, weave, body movement, etc.), and the vapour resistance of the air R_A (this will depend greatly on the air movement). Finally we have the equation

$$\frac{E}{S} = \frac{P_S - P_A}{R_{cl} + R_A} \dots\dots\dots\dots\dots\dots\dots\dots 15$$

The reader may feel that unnecessary space is being given to discussion of evaporation of moisture in a book dealing with man in the cold, and that we may confidently take the 'wetted area' of man in the cold as always at the minimum 10 per cent value for insensible perspiration. This is an error, which has had serious practical consequences, for two reasons. The first is that muscular exercise brings about sweating in man even in very cold environments, though the skin temperature at which it appears may be much lower (10). Indeed those experienced in living in the cold are emphatic that sweating in the cold is the cause of much of the discomfort and difficulties in present Arctic clothing (see later chapter). Clothing is seldom so quickly adaptable that it can be adequate for men at rest in the cold and yet reduced enough to prevent overheating and sweating when there is heavy exercise. The second reason why sweating must not be neglected in the cold, particularly in problems of protection of military personnel, is that sweating occurs not only due to thermal stimulation but also, independent of temperature conditions, in response to emotional stimuli. Indeed the two physiological functions of sweating are clearly separated. Emotional sweating takes place almost exclusively on the palms of the hands, axillae, the soles of the feet and the forehead (10), while thermal sweat glands are widely distributed. In conditions of stress, or even of 'attention', and certainly in anxiety or fear, sweating from hands and feet will be profuse, even in the coldest conditions. Neglect of this factor in designing hand gear or foot gear for military personnel may lead to serious consequences.

4. Units of Thermal Insulation and of Vapour Resistance
The Clo Unit

We are, of course, at liberty to use any combination of units of heat, temperature, area and time in the fundamental equations, so long as they are consistent with each other. Much valuable cross-fertilization amongst those interested in the problems of human heat exchanges with the environment, but in different scientific disciplines, has been impossible because the different groups used quite different systems of units. The heating and ventilating engineers habitually use, and think in terms of, British Thermal Units (B.T.U.) of heat (or even such units as 'lbs' of refrigeration) and they use the Fahrenheit scale of temperature, square feet for area and usually hours or minutes for time. On the other hand, physicists prefer the C.G.S. system of units, i.e. calories, °C, square centimetres and seconds. Among metereologists and climatologists, some use the Fahrenheit scale, others the Centigrade, with a noticeable swing towards the latter of late years. Physiologists use a curious mixture, namely kilocalories, °F or °C impartially, square metres and hours. Consequently there could be a corresponding number of different units for thermal insulation.

For the physicist, these would be in °C divided by calories/sq. cm/sec (or possibly watts/ sq. cm), for the engineer, °F divided by B.T.U./sq. ft./min.

When in addition to the professional groups, non-technical persons such as generals or admirals sitting on war-time committees had to understand the results of work in this field, it seemed to a group of workers in three different laboratories (11) that there would be a usefulness in introducing a new unit for thermal insulation, particularly in connexion with clothing. The new unit could be translated into whatever system of established units was preferred, merely by the use of a numerical constant, but might justify its existence by carrying a concrete meaning to the layman. Such a unit is the 'clo' unit of thermal insulation, and it has received wide adoption.

The fundamental units in which the clo unit is expressed are chosen so as to be familiar to the physiologist. Obviously heat should be in kcal/sq.m/hr. The magnitude of the clo unit of insulation was then chosen to approximate that of a familiar garb, the normal indoor clothing worn by sedentary workers in comfortable indoor surroundings, i.e. a 'business suit' and the usual undergarments. Most people in cities wear 1 clo unit of thermal insulation most of the time. Europeans wear habitually more than 1 clo unit, perhaps up to 1·3 clo units, while Americans wear less than 1 clo unit, perhaps down to 0·7 clo units. The difference is due to the difference in the usual indoor temperatures. 'Room temperature' in the United States means about 24°C (75°F), in Britain about 18·3°C (65°F). Examination of one Russian paper showed that 'room temperature' meant 12°C (54°F).

Physiological information tells us that a resting sitting man produces about 50 kcal/sq.m/hr, and therefore if his clothing is to keep him comfortable in normal indoor environments, 21°C (70°F), with a low air movement of 20 ft./min, or 10 cm/sec, and relative humidity less than 50 per cent, his total heat loss must also be 50 kcal/sq.m/hr. Subtracting 25 per cent for the evaporative loss, this is 38 kcal/sq.m/hr through his clothing by means other than evaporative loss. Also physiological experiments have shown that the average skin temperature for comfort, in a number of individuals, is about 33°C (91·4°F), so we use this in equation 8. Inserting these values we have

$$I_A + I_{cl} = \frac{33 - 21}{38} = 0·32 \; \frac{°C}{\text{kcal/sq.m/hr}}$$

Previous work at the Pierce Laboratory by Winslow, Gagge and Herrington (12) has shown that the insulation of the air for the human body under these conditions of air movement will be

$$I_A = 0.14 \; \frac{°C}{\text{kcal/sq.m/hr}}$$

We therefore must have an insulation of the clothing, which we wish to define as 1 clo unit, $1_{cl} = 0.32 - 0.14 \; \dfrac{°C}{\text{kcal/sq.m/hr}}$

The clo unit is therefore *defined* as $0.18 \dfrac{°C}{\text{kcal/sq.m/hr}}$

$$\text{or } 0.88 \; \frac{°F}{\text{B.T.U./hr/sq. ft.}}$$

The *definition* of the clo unit is in terms of the fundamental units and the appropriate constant (0·18 in the units most often used by physiologists). The *visualization* of the clo unit is in terms of the following statement: '1 clo unit of thermal insulation will maintain a resting-sitting man, whose metabolism is 50 kcal/sq. m/ hr, indefinitely comfortable in an environment of 21°C (70°F), relative humidity less than 50 per cent, and air movement 20 ft./min.'

There has been some criticism of the clo unit, stating that it cannot be a scientific unit, since it is based on experimentally determined and variable values (as of the heat production of a sitting-resting man). This criticism is based on a misconception that the unit is *defined* by this *example*, rather than by the numerical constant translating it into the standard units of physics. The numerical factor was chosen, and the example given, so as to make it easy to visualize the clo unit. Further experiments cannot alter the basis of the clo unit, though it might be decided that the numerical factor in the definition would give a better visualization of the meaning of 1 clo unit, if it were altered slightly. For example, in post-war Britain it might better be altered to represent the insulation of clothing that keeps a man comfortable at 15·5°C (60°F).

The Unit of Vapour Resistance

Just as the clo unit is defined from the equation of heat flow, (equation 7), so the unit of vapour resistance is defined from the equation of evaporative heat loss (equation 14). However, since there is very little uniformity apparent in the vapour resistance of the various clothing worn under usual circumstances, there would be no point in choosing the definition in the hope of attaining easy visualization of what one unit of vapour resistance meant. We would no doubt call such a unit the 'clammy unit' or perhaps the 'Fug'. Instead the workers who pioneered in this field (13) use the vapour resistance of 1 cm of 'dead air' (air which has no convection currents in it by reason of being confined in a very narrow space or immobilized by a 'filler' of fibres in an air space). Fourt and Harris (14) who took up the subject later also adopted this device. Therefore the definition of vapour resistance is:

One unit of Vapour Resistance = 1 cm of dead air =

$$8 \cdot 6 \; \frac{\times \; \text{mm Hg difference of vapour pressure}}{\text{grams water vapour diffusing}}$$

The numerical factor is the result of experimental determinations of the transmission of vapour by dead air spaces, which agree with the accepted values for vapour diffusion coefficient in air (15). For the purposes of calculating the heat exchanges, we have to introduce the factor 0·60 kilocalories per grams of water evaporated, and equation 14 becomes

$$\frac{E}{S} = (5 \cdot 0) \frac{P_S - P_A}{R_{cl} + R_A} \text{kcal/sq.m/hr} \quad \ldots\ldots\ldots 16$$

where R_{cl} and R_A are in cms of dead air.

5. The Heat Production

The factors affecting the heat production of an animal will be discussed in some detail in a later chapter. Here it is sufficient to point out that physiologists already are accustomed to reckon this per unit area of the surface of the body, i.e. in kcal/sq.m/hr so that it is in the form required for the left hand side of the equation of the thermal steady state. Rate of heat production is calculated from a measurement of the rate of oxygen consumption, a knowledge of the type of food which is being burned, from the respiratory quotient, and the measured value of kilocalories per litre of oxygen consumed for this type of foodstuff (see book by Lusk— *The Science of Nutrition*). An important new contribution is that of Weir (16) who has demonstrated that the rate of oxygen consumption and of the total ventilation alone gives as good a calculation of heat production, possibly a better one, as the more elaborate methods using the R.Q. It is well known that the thermal equivalent of a litre of oxygen absorbed varies considerably according to the R.Q. Weir shows that the calories per litre of expired air, in contrast, vary very little with the R.Q. With very little error indeed, we can use a value ($1 \cdot 046 - 0 \cdot 05 \ O_e$) for the calories per litre of expired air, where O_e is the percentage of oxygen in expired air. For details the original paper should be consulted. This way of calculating heat production lends itself to the convenient experimental method of the continuous measurement of oxygen percentage by the Pauling oxygen meter. Throughout this book the heat production will be denoted by M in kcal/sq. m/hr.

6. The Heat Debt

If in any period heat production and heat loss are not equal, the difference will change the average temperature of the tissues of the body, by the equation

$$M - H = \frac{m \ s}{S} \times \frac{d\theta}{dt} = D \dots\dots\dots\dots\dots 17$$

where M and H are the heat production and heat loss respectively, in kcal/sq. m/hr, m is the mass of the body, s its specific heat, and S is the surface area of the body, θ is the average temperature of the body in °C. The rate of change of average body temperature, will be, from this equation, in °C per hour. Because of the familiarity to physiologists of the 'oxygen debt', which accumulates when the rate of oxygen absorption by the lungs is not as great as the rate of oxygen utilized by the body, it is convenient to call the right hand side of the equation, or rather *minus* this, the heat debt, D (or rate of heat debt) in kcal/sq. m/hr. The heat debt will be positive when heat loss exceeds the heat production, negative when the heat production exceeds heat loss. This seems a more generally understood convention, than the one originally introduced by Gagge and his co-workers (17), who used the term 'heat storage' as positive when the heat loss exceeded heat production and the average body temperature was falling. It was difficult to see where heat was 'stored' on this convention.

Calculation of the heat debt from measurements of temperature at available points in the body is of great importance in 'indirect' animal calorimetry, for it enables the heat loss to be estimated from the measurements of heat production by the above equation. Accurate calculation of the heat debt, even when valid measurements of the temperatures of the surface of the body, and of the 'core' temperature are available, is a difficult matter. Originally calorimetrists used only the changes in 'core' temperature, as given by deep rectal temperatures, as an indication of the change of average body temperature. It became obvious, when direct and indirect calorimetry were done simultaneously, that the discrepancies between the measured heat production and the measured heat loss could not be adequately explained in terms of changes in rectal temperature only. Indeed cases often occurred where the heat loss had exceeded the heat production, so that average body temperature must have fallen, yet the rectal temperature rose.

The next approximation (18) was to use in addition to the rectal temperature, representing the 'core' temperature, an average surface temperature and so attempt to obtain a better average for the temperature of the tissues. How were the rectal temperature and this average surface temperature to be 'mixed'? This was attacked in two ways. First a theoretical calculation of the 'mixing coefficient' was made, based on the experimental observations of Bazett and McGlone (19) on how the temperature rose at different depths below the skin (i.e. the nature of the internal temperature gradients), on data as to relative weights of different appendages of the body (20), and on mathematical use of these with crude assumptions as to the distribution of heat production throughout the tissues. The result was an equation for the average temperature of the tissues, as the next approximation to using rectal temperature alone.

$$\theta = 0 \cdot 64 \; T_R + 0 \cdot 36 \; T_S \; \dots\dots\dots\dots\dots 18$$

where T_R and T_S are the rectal and average skin temperatures. The latter is calculated from a number of temperatures at different strategic points on the body surface. For details the reader is referred to the original paper (18).

The empirical method was then used. In 40 periods of human calorimetry in which heat production and heat loss were simultaneously measured, so that their difference could be calculated, the best value of 'mixing coefficient' a in an equation of this type

$$\theta = a. \; T_R + (1 - a) \; T_S \; \dots\dots\dots\dots\dots 19$$

was calculated by statistical methods. By 'best value' is meant the value which statistically gives the best agreement of the heat debt calculated from body temperatures, and the measured heat debt from the difference between heat production and heat loss. For

D

these 40 periods, on several subjects, the best value of a was 0·70 and the best mixing equation

$$\theta = 0·7\ T_R + 0·3\ T_S \quad \dotsc\dotsc\dotsc\dotsc\dotsc\dotsc\dotsc 20$$

Obviously the calculation of heat debt, using this empirically determined equation, must still be very rough. The mixing coefficient would certainly differ from subject to subject according to their body build and distribution of fat, and internal gradients of temperature must alter greatly in form with vasomotor adjustments. Yet the use of the average skin temperature as well as the rectal temperature does markedly reduce the discrepancies obtained with the use of rectal temperature alone, as shown in the following table, for these 40 periods.

	Using Rectal temp. alone	*Using rectal and skin temp. by formula*
Average discrepancy between calculated and measured heat debts.	7·5%	5·5%
No. of periods with discrepancy more than 10%	10 out of 40	4 out of 40
No. of periods with discrepancy less than 5%	16 out of 40	25 out of 40

We may conclude that, with this rather small number of subjects and small range of conditions, the heat debt can be calculated by the aid of such a 'mixing formula', from rectal and average surface temperatures, probably within 5 per cent of the true value; part of the discrepancies must be due to errors in measurement of heat production.

Hardy and DuBois (21) prefer to use a different mixture, namely $0·8\ T_R + 0·2\ T_S$. It is unlikely that the choice can be narrowed down unless much more extensive measurements are made. Nor is it likely that to do so would be very useful, since the variability of the true mixing coefficient from subject to subject and with different physiological states must be great. An illustration is given in Fig. 8 of how little difference it makes, what coefficient is used, between the wide limits for a of 0·8 and 0·6.

It must not be thought, since the coefficient multiplying the change in rectal temperature is about twice that of the change in surface temperature, that the rectal temperature is correspondingly more important in the calculation of heat debts. On the contrary, the surface temperature is physiologically so much more variable than the rectal temperature (changes being often four times as great) that in the calculation it is usually the change in surface temperature that contributes more to the change in average temperature. Indeed about as good a calculation of heat debt could

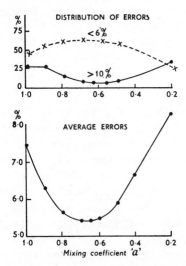

FIG. 8. 'Errors' re discrepancies between direct and indirect measurement of human heat exchanges in 40 calorimetry periods, when different 'mixing coefficients' are used in deriving the average temperature of the body tissues. (From Murlin and Burton. *J. Nutr.*, **9**, 272, 1935.)

be measured using the changes in surface temperature alone, by the equation:

$$\triangle \theta = 0.61 \triangle T_S$$

as by using the rectal temperature alone in the equation:

$$\triangle \theta = \triangle T_R$$

Using both rectal and surface temperature, however, is much better than either of these.

7. The Specific Heat of the Body

In the experimental determination of the best 'mixing coefficient', described above, the value 0.83 was assumed for the average specific heat of the body tissues. This value was based upon an average of the specific heats of the various tissues, measured 'in vitro' many years ago (22). It was possible, from the data of the

40 periods of calorimetry, to make an estimate of the value of specific heat which gave the least average discrepancy. This turned out to be also 0·83, but the striking feature is how little difference it makes what value, in the range from 0·7 to 0·9, is used, as shown in the table below.

Value of specific heat assumed	Average discrepancy in calculated and measured heat debt (per cent)
1·0	5·87
0·90	5·55
0·83	5·52
0·70	5·56
0·60	5·59

This may be regarded as the only 'in vivo' determination of the average specific heat that has been made for the whole human body. Recently (23) an experimental determination of the specific heat of fingers, with the circulation occluded, has been made. The result for the specific heat was 0·75. We might well continue to use the accepted value of 0·8 in calculations of heat debt, realizing how, like the 'mixing coefficient', it must vary greatly between individuals and even in different physiological states.

One other direct determination of the specific heat of the whole body has been made, to our knowledge. Hart (24) measured the heat given up by 15 dead white mice, after heating them in a water bath to a known temperature (presumably this would be uniform throughout), and dropping them into a water calorimeter in the standard manner. The average value was 0·824 ± ·005.

8. Loss of Heat by Warming Inspired Air

In most discussions of the heat loss of the human body to be found in textbooks of physiology, the loss of heat in raising the inspired air to body temperature is listed as one item. It can be admitted as playing a part in the heat balance though it is a small one. Air has a very low specific heat per gram, and if the kilo-calories required to raise inspired air to body temperature is esti-mated, the result will be found to be astonishingly small.

An average man at rest, with a surface area of 1·8 sq. m, will have a total lung ventilation of about 8 litres/min., or 480 litres/hour. This air, inspired at the environmental temperature, say 20°C (68°F), is raised to 37°C (98·6°F) in the lungs, i.e. by 17°C (30·6°F). The specific heat of air, at normal atmospheric pressure, is 0·24 calories per gram per °C. 480 litres of air will have a mass of 576 g, density of air = ·0012 g/ml, and to raise this mass of air by 17°C (30·6°F) will require 2,350 calories, or 2·35 kcal. This represents a loss of heat to the man of only 2·35/1·8 or 1·3 kcal/sq. m/hr. This is only about 2 per cent of the total heat loss of a resting man (50 kcal/sq. m/hr.) The loss of heat in saturating the inspired air with water vapour (about 5 or 6 kcal/sq. m/hr), i.e. the heat loss by evaporation from the respiratory tract, is considerably greater.

In heavy exercise the ventilation rate will increase greatly, but only in proportion to the increased heat production and heat loss. The *proportional* loss by the warming of inspired air therefore does not increase in exercise.

In very cold environments, where air is inspired at, say, − 40°C (− 40°F), the heat required to warm it to body temperature, i.e. by 87°C (157°F) will begin to be of more importance, but will still be only 9 per cent of the resting heat production.

9. Loss of Heat by Urine and Faeces

This also is listed as an item of heat loss of the human body in most textbooks. Its magnitude is even less than that of warming inspired air, as it seldom will be more than 1 per cent of the total resting heat loss. Here, however, there may be considerable doubt, whether this item should be admitted at all in the heat balance.

It is true that the urine and faeces leave the body at approximately the core temperature, and subsequently cool to the temperature of their surroundings, in so doing giving up heat to the environment. But the expulsion of excreta from the body does not lower in any way the body temperature. Heat is lost, but a corresponding mass of the body is also lost with the excreta. It does not require any increase in heat production, or decrease in heat loss by other routes to make up for this heat 'lost' in the excreta, and the thermal state of the body is unchanged, unless it be in that the man is a little lighter and has a slightly smaller surface area as well.

It is quite otherwise when water is lost from the body by evaporation. Here for every gram of water vaporized the body must supply 0·60 kcal of heat production. There is also the heat involved in cooling of the water vapour, so formed, to environmental temperature, and in addition the work done in expanding it to the relative humidity of the environmental air. These items have been taken into account in the figure 0·60.

It is to be concluded, therefore that, whether or not the heat losses in excreta are to be considered is very debatable. Probably they have no place in our heat balance equation. Fortunately the outcome of the argument will have no significant practical effect.

10. The Complete Equation of Heat Balance

Whether a thermal steady state is achieved or not, the equation of heat exchange of the animal body is:

$$M - D = H + E \quad \dots \dots \dots \dots \dots \dots \dots 21$$

where

$$H = 5 \cdot 5 \; \frac{T_s - T_A}{I_{cl} + I_A} \text{ and } E = 5 \cdot 0 \; \frac{P_s - P_A}{R_{cl} + R_A}$$

In this equation, thermal insulations are in clo units, vapour resistances in cm of dead air, H and E in kcal/sq. m/hr. When a thermal steady state is achieved, D is zero.

This is a simple, self-consistent scheme for predicting requirements for protection in cold climates, and the limits of tolerance of man and animals when the limits of such protection are known. In a later chapter it is shown how this equation gives us a consistent way of estimating the total thermal stress of a given environment on man. In the next few chapters the various insulations will be considered separately, before the equation is used as a basis for examining the means available to an animal for maintaining heat balance, and the effectiveness of these.

All this would be much more satisfactory if we did not have evidence that even in experiments where the physiological state of a man, and the environmental conditions, were well controlled to approximate constancy, *the steady state may not be reached even after several hours.* In actual circumstances of normal life, and especially in military life, where periods of activity and rest follow each other in rapid succession and environmental factors change from time to time, it is very doubtful whether we are ever dealing with even an approximation to the steady state. Equation 21 still applies, but the heat debt D becomes of relatively great importance. It is therefore most unfortunate that we cannot do better than make a rough approximation to its calculation for the body itself. Where clothing is concerned, we are at present very much at sea as to how to calculate the heat debt at all.

REFERENCES

1. HARDY, J. D. Heat Transfer. Chapter 3 in 'Physiology of Heat Regulation.' L. H. Newburgh, Saunders, Philadelphia, 1949.
2. BEDFORD, T. 'Basic Principles of Ventilation and Heating.' Lewis, London, 1948.
3. BURTON, A. C. The Application of the Theory of Heat Flow to the Study of Energy Metabolism. *J. Nutrition*, **7**, 481, 1934.
4. DuBOIS, E. F. 'Basil Metabolism in Health and Disease.' Lea and Febiger, Philadelphia, 1927.
5. GAGGE, A. P. A New Physiological Variable associated with Sensible and Insensible Perspiration. *Am. J. Physiol.*, **120**, 277, 1937.
6. ARMSTRONG, H. and BURTON, A. C. Unpublished Observations.
7. CHRISTIE, R. V. and LOOMIS, A. L. The Pressure of Aqueous Vapour in the Alveolar Air. *J. Physiol.*, **77**, 35, 1932.
8. McCUTCHAN, J. W. and TAYLOR, C. L. Respiratory Heat Exchanges with Varying Temperature and Humidity of Inspired Air. *J. Appl. Physiol.*, **4**, 121, 1951.
9. MOLE, R. H. The Relative Humidity of the Skin. *J. Physiol.*, **107**, 399, 1948.
10. KUNO, Y. 'The Physiology of Human Perspiration.' Churchill, London, 1934.
11. GAGGE, A. P., BURTON, A. C. and BAZETT, H. C. A Practical System of Units for the Description of the Heat Exchange of Man with his Environment. *Science*, **94**, 2445, 428, 1941.
12. WINSLOW, C-E. A., GAGGE, A. P. and HERRINGTON, L. P. Heat Exchange and Regulation in Radiant Environments Above and Below Air Temperature. *Am. J. Physiol.*, **131**, 79, 1940.
13. TUCKER, J., GOODINGS, A. C. and KITCHING, J. A. The Permeability of Textile Materials to Water Vapour. Associate Committee on Aviation Medical Research, Report No. C. 2655, 1944.
14. FOURT, L. and HARRIS, M. The Resistance of Fabrics to the Diffusion of Water Vapour. *Textile Research. J.*, **17**, 256, 1947.
15. GREGORY, J. The Transfer of Moisture Through Fibres. *J. Textile Inst. Trans.*, **21**, 66, 1930.
16. WEIR, J. B. de V. New Methods of Calculating Metabolic Rates with Special Reference to Protein Metabolism. *J. Physiol.*, **109**, 1, 1949.
17. GAGGE, A. P., WINSLOW, C-E. A. and HERRINGTON, L. P. The Influence of Clothing on the Physiological Reactions of the Human Body to Varying Environmental Temperatures. *Am. J. Physiol.*, **124**, 30, 1938.
18. BURTON, A. C. The Average Temperature of the Tissues of the Body. *J. Nutrition*, **9**, 261, 1935.
19. BAZETT, H. C. and McGLONE, B. Temperature Gradients in the Tissues in Man. *Am. J. Physiol.*, **82**, 415, 1927.
20. BRAUNE, W. and FISCHER, O. 'Determination of the Moment of Inertia of the Human Body and the Limbs.' S. Hirse 2, Leipzig, 1892.
21. HARDY, J. D. and DuBOIS, E. F. Basal Metabolism, Radiation, Convection and Evaporation at Temperatures of 22° to 35°C. *J. Nutrition*, **15**, 477, 1938.
22. PEMBREY, M. S. 'Animal Heat', in 'Textbook of Physiology.' E. A. Schafer. Vol. I, p. 838. Hodder and Stoughton, London, 1898.

23. GREENFIELD, A. D. M., SHEPHERD, J. T. and WHELAN, R. F. The Average Internal Temperature of Fingers Immersed in Cold Water. *Clin. Sc.*, **9**, 349, 1950.
24. HART, J. S. Calorimetric Determination of Average Body Temperature of Small Laboratory Animals and its Variation with Environmental Conditions. *Canad. J. Zoology*, **29**, 224, 1951.

CHAPTER 3

THE THERMAL INSULATION OF THE AIR

The Insulation of the Air, I_A

This is defined by the equation:

$$H = \frac{T_{cl} - T_A}{I_A} \text{ or } I_A = \frac{T_{cl} - T_A}{H} \quad \ldots\ldots\ldots 1$$

In words; the insulation of the air is obtained by dividing the temperature gradient from the surface of the clothing to the environment by the non-evaporative heat flux through the 'surface skin' of air, as some have called it.

It is known that the heat loss of a warm surface to an environment of air is by two main routes, excluding loss by evaporation. These are *convection*, in which air is brought in contact with the surface either by forced or natural currents of air (in 'natural convection' the air currents are due to the thermal gradient only), and this air takes up the temperature of the surface, and carries it to distant points. The second means of transport of heat is by *radiation*, which is an electro-magnetic disturbance emitted by the warm surface, and which would take place unchanged if the air were replaced by a vacuum. Air transmits this radiation over long distances but it is eventually absorbed by the molecules of air, so that the air in the environment tends to rise in temperature. Radiant heat loss is thus independent of the air movement. In contrast, convective heat loss is very greatly increased when air movement is greater, as in a wind. A third mode of transport of heat into the environment is by pure conduction. By the conductivity of a gas, a physicist means the power to transport heat from molecule to molecule by the collisions of the molecules with each other, in which they exchange their kinetic energies. The temperature of a gas is nothing but the measure of this molecular kinetic energy. This exchange of energy by collision transports heat from one part of the gas to another. In the process of pure conduction no 'drift' of molecules themselves from one region to another is supposed, in contrast to convection, and in actuality this condition is never obtained, for if temperature gradients exist there will be such drifts due to the differences of density that arise. As will

47

be seen, conduction in air is of very little practical importance. A great deal of confusion has resulted from lack of appreciation of the meaning of 'pure conduction'.

The well-informed student knows the laws that govern the transport of heat by convection and by radiation, and this often raises a difficulty in his mind as to how our linear law based on Newton's law of cooling can apply to the total heat loss. For though the empirical laws of convection are approximately linear, i.e. the heat loss is proportional to the difference of temperature raised to some power not very different from unity (1), the law for radiation is very far from linear. The intensity of radiation from a unit area of surface is proportional to the fourth power of its absolute temperature (Stefan's law). Taking into account the fact that radiation is being received by the surface from the environment, the net heat loss is given by

$$H = E \, \sigma \left\{ T_{cl}^4 - T_A^4 \right\} \quad \dots\dots\dots\dots\dots\dots\dots 2$$

E is a constant dependent on the nature of the surface which expresses how closely it approximates to the perfect radiator—'perfect black body'. E is called the 'emissivity' and is normally expressed as a percentage: a 'black body' has an emissivity of 100 per cent; σ is Stefan's universal constant. This law has a very different status from the empirical laws of convection, or of the heat loss, because it is based on thermodynamics and can be proved without regard to theories of the mechanism of radiative exchange.

How then can the total heat loss be at all proportional to the difference of temperature, in view of this fourth power law? The answer is that algebraically:

$$T_{cl}^4 - T_A^4 = \left\{ T_{cl} - T_A \right\}$$
$$\left\{ T_{cl}^3 + T_{cl}^2 \times T_A + T_{cl} \times T_A^2 + T_A^3 \right\} \quad \dots\dots\dots\dots 3$$

The exchange of radiation will be proportional to the difference of temperature $(T_{cl} - T_A)$ to the extent that, over the working range, the second term on the right-hand side is a constant. Now the maximum working range is found to be a temperature difference of about 25°C (77°F) between the surface of clothing and the environment; a greater difference would result in a heat loss more than three times the resting metabolic rate so a steady state would be impossible. In the usual environments the absolute temperature is about 300°K (27°C, 80°F). Taking T_A as 20°C or 293°K, and T_{cl} as 10° higher (303°K, 30°C), the differences of the fourth powers is $1\cdot17 \times 10^9$. For a 20° difference $(T_{cl} = 40°C$ or 313°K) the difference of the fourth powers is $2\cdot23 \times 10^9$. Now $\dfrac{2\cdot23}{1\cdot17}$ is $2\cdot09$, i.e. the departure from linearity, which would give exactly twice the heat loss for twice the difference of temperature, amounts to only 5 per cent even over this rather unusually great range of temperature. We must remember that only about half of the total heat loss is due to radiation, the rest to convection (the fraction of course depends on the air movement) so that the departure from linearity for the total heat loss is even less.

This explanation of the validity of Newton's law was satisfactory until, during the last war, we became interested in the heat exchanges of man in very cold surroundings, say down to $-40°C$ ($-40°F$). The calculation above shows that the proportionality constant of Newton's law ($1/I_A$) will be approximately unchanged over the range of differences of temperature ($T_{cl} - T_A$), but will it be the same constant when the absolute level of both temperatures is so different? For example, while, as above, the difference of the fourth powers of absolute temperature for a 10°C difference ($T_{cl} - T_A$) was $1\cdot17 \times 10^9$ when T_A was 20°C (293°K), the difference of fourth powers for a 10° difference when $T_A = -40°C$, is only $0\cdot54 \times 10^9$. This means that the radiative exchange for the same 10° difference between surface and air is less than half what it was at higher temperature, and we might assume that a very different constant I_A would have to be used at the low temperature. Yet experiment showed that the total heat loss, by radiation plus convection, still followed with close approximation the same law with the same value for I_A as had been determined for the environments of higher temperature.

Investigation revealed that this was due to what can only be regarded as a most fortunate accident of physics (2). Although in colder environments the loss by radiation decreases, the loss by convection, for the same velocity of air movement, increases because of the increased density, which gives the air an increased power to carry heat as its temperature falls. The increase happens to compensate, very nearly, the decrease in radiative loss, and this is so over practically the whole range of values of air movement.

Results of measurements at the Pierce Laboratory at Yale (3) have given for the total heat loss of the human body, the following equation:

$$I_A = \frac{1}{0\cdot61 + 0\cdot19V^{\frac{1}{2}}} \text{ clo units} \qquad \dots\dots\dots\dots 4$$

where V is the velocity of air movement in cm/sec.

The second term $(0 \cdot 19 V^{\frac{1}{2}})$ represents the loss of heat by convection, which depends on the square root of the velocity of air movement. The term $0 \cdot 61$ represents the loss of heat by radiation. The experiments on which this equation is based were made at the usual room temperatures (25°C (77°F) or 298°K). We can now correct the equation for change of temperature. The term connected with radiation loss must be multiplied by the factor $\left(\dfrac{T}{298}\right)^3$ since calculus (from equation 3) indicates this. The term for the convection loss must be corrected for the change in density, which varies inversely with the absolute temperature, i.e. must be multiplied by the factor $\left(\dfrac{298}{T}\right)$. We have then, for all temperatures:

$$I_A = \frac{1}{0 \cdot 61 \times \left(\dfrac{T}{298}\right)^3 + 0 \cdot 19 V^{\frac{1}{2}} \times \left(\dfrac{298}{T}\right)} \qquad \ldots \ldots \ldots \ldots 5$$

The results for a variety of velocities of air movement, at 25°C (77°F) and -40°C (-40°F) are shown in the following table, which also gives the partition of the total heat loss between radiation and convection.

Air Movement			Room temp. 25°C (77°F)			−40°C (−40°F)		
cm/sec	miles/hr	ft./min	% Rad'n.	% Convection	Insulation clo	% Rad'n.	% Convection	Insulation clo
9	0·21	18	52	48	0·85	32	68	1·08
25	0·56	49	39	61	0·64	22	78	0·75
36	0·81	71	35	65	0·57	19	81	0·64
49	1·09	96	31	69	0·52	16	84	0·57
81	1·81	159	26	74	0·43	13	87	0·46
121	2·69	238	23	77	0·37	11	89	0·39
225	5·04	443	18	82	0·29	8	92	0·29
400	9·00	790	14	86	0·23	6	94	0·22
625	14·0	1230	11	89	0·19	5	95	0·18
1024	22·9	2010	9	91	0·15	4	96	0·14
2500	56·0	4920	6	94	0·10	3	97	0·09

It will be apparent from the table that although the partition of losses alters very greatly indeed with temperature, the total is approximately the same at these two temperatures. For example, at a velocity of 5 miles/hr in an environment of 25°C (77°F), radiation carries 18 per cent and convection 82 per cent of the total loss, whereas at -40°C (-40°F) radiation is only 8 per cent and convection 92 per cent of the total. Yet the total heat losses (per °C difference) are identical. Fig. 9 illustrates this.

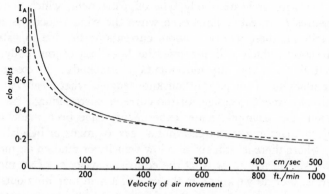

Fig. 9. Heat loss at different air velocities. The heavy line is for an environmental temperature of —40°C (—40°F), while the broken line is for 25°C (77°F).

Standard Values of IA at Different Wind Velocities

This fortunate circumstance allows us to adopt a standard set of values for the insulation of the air, I_A, at different velocities of air movement, given in the table below:

STANDARD VALUES FOR I_A

Insulation clo units	0·1	0·15	0·20	0·30	0·40	0·50	0·60	0·70	0·80	0·85
Velocity of air movement miles/hr	51·0	22·7	11·9	4·85	2·39	1·37	0·85	0·57	0·40	0·34
ft./min	4,500	2,000	1,050	425	210	120	75	50	35	30
cm./sec	2,280	1,015	534	216	107	61	38	25	18	15

As the graph shows, these will be independent of the temperature to within 0·05 clo units.

The shape of the graph indicates that while at very low air movements the degree of air movement makes a great deal of difference to the heat loss of a man, the decrease in I_A is very little more with a wind of hurricane force than with one of 10 miles/hr. It must not be concluded that this applies in practice to the clothed man, for the wind may penetrate the clothing and produce an air movement there which will greatly affect its thermal insulation, I (see page 64). Thus in practice the heat loss of a clothed man may be very much greater in a wind of 50 miles/hr than in one of 10 miles/hr.

By the velocity of air movement is meant the 'random' velocity rather than the 'drift velocity', which would be measured by a vane type of anemometer. Such random velocities are best measured by the hot wire or heated-body type of instrument, which actually measures I_A directly. Thus even when the wind velocity is zero, i.e. still air, there will be random currents of air, due to natural convection, which will increase the heat loss of the body. In practical cases, too, the movements of the body will create air currents. One of the gaps in our knowledge is what mean velocity of air movement to assign for the currents surrounding a man in motion. For example, many experiments have been made with subjects marching on a tread-mill at, say, $3\frac{1}{2}$ miles/hr in a laboratory where there is 'still air' or a low velocity of random air movement. We do not know what the equivalent velocity of air movement is. Nor do we know, in the case of a man marching outside, how to add such an equivalent velocity due to his movements, to his velocity of progression and the wind velocity, to obtain the resultant air movement.

Effect of Altitude

One further modifying effect requires discussion, namely that of altitude on thermal insulation. Since this is due to the dead air entrapped, an effect would be expected because of its change in density (as already introduced in discussing the effect of temperature on the insulating power). The term representing convective heat loss must be multiplied by the square root of the ratio of density at altitude to that at ground level (4). As a result, for still conditions, where I_A is 0·8 clo units at ground level, the insulation of the air would rise to 1·1 clo units at 20,000 ft. This may be of little importance in the total insulation of an airman, but the insulation of his clothing, which may be as much as 5 clo units, will be correspondingly increased (to 7 clo units). This has been tested experimentally with flying clothing on an 'artificial man' in an altitude chamber, giving the graph, Fig. 10. In addition 'insolation', radiation from the sun, will greatly increase at altitude and effectively lower the thermal demand (Chapter 7).

Transfer of Heat in 'Still Air'

Insight into the nature of the layer of 'still air' surrounding the body and the transfer of heat through it is given by experiments on the heat flow through narrow air spaces between two blackened horizontal plates (5). Fig. 11 shows the results. The insulation at first increases linearly with the thickness of the air space, but the slope steadily decreases until after a thickness of about $\frac{1}{2}$ in. is

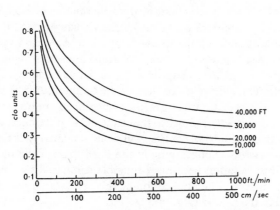

FIG. 10. Insulating power of air at various air movements and altitudes. (Redrawn by Belding from Burton. *Physiology of Heat Regulation*, ed. Newburgh.)

reached there is no more gain in insulation by making the space wider. In fact, the insulation for wider spaces actually decreases (Fig. 11). This is due to the development of convection currents more and more freely as the space becomes wider. When the plates of the apparatus are made reflecting for radiation (emissivity about 5 per cent) instead of black (emissivity 100 per cent), a curve of the same shape is obtained, but because the transfer by radiation is now almost negligible, the ordinates are increased by about 40 per cent. Narrow empty air spaces separated by reflecting

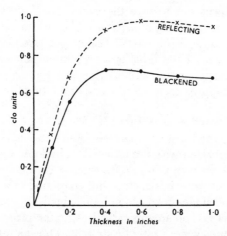

FIG. 11. Insulation of an empty air space between two plates with thickness of the space. The apparatus lacked a guard-ring so results are approximate only.

partitions are thus an excellent form of thermal insulation, and are so used in some modern refrigerators. Obviously, there is no useful gain in making such air spaces more than ½ in. Actually in practice, there is no need to use reflecting partitions if a series of successive very narrow spaces are used. Since the radiative exchange depends on the difference of fourth powers of the temperatures of adjacent partitions, the radiation will greatly decrease. In contrast the convection and conduction, with the same total difference of temperature across the whole space, will change relatively little, if the spaces are narrow enough to be in the linear part of the insulation curve. Thus radiation becomes negligible compared to convection in narrow spaces. The application of this to clothing insulation is discussed in the next paragraph.

The Insulation of a 'Dead Air' Space

Evidently air enclosed in a very narrow space is greatly superior as thermal insulation to 'still air' in a wider space or unconfined, because of the prevention of currents of natural convection. It is proposed that air in which these are completely prevented should be called 'dead air' in distinction to 'still air'.

The thermal insulation of which 'dead air' is capable is given by the initial slope of the curve (Fig. 11), i.e. the clo units per inch of air in an infinitely thin air space. Since, as explained, radiative exchange can be made negligible in successive very narrow spaces, we should use the slope of the curve obtained with the reflecting faces. This turns out to give the value of 4·7 *clo units per inch thickness for the insulation of dead air*. This would correspond to a conductivity value K of ·00083 kcal/sq. cm/sec./°C/cm or 0·24 B.T.U./sq. ft./hr/°F/in. This is a very important constant, as it turns out that it applies also to the best thermal insulation of clothing, and approximately to the best-insulating fur of animals.

The Insulation of a Filled Air Space

When the experiments with air enclosed between two horizontal plates were repeated with the space filled by loose material of low bulk density (less than 5 lb./cu. ft.), such as cotton-wool, the curve started with the same initial slope but continued in a straight line, with this same slope of 4·7 clo units per inch thickness (Fig. 12). Data was obtained on an apparatus lacking the proper guard ring. If corrected the curves would be straight. Thus an insulation of as many clo units as required could be obtained by having a 'filled'

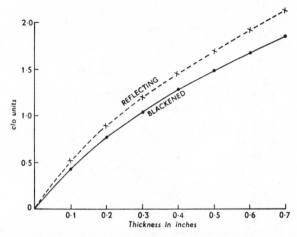

Fig. 12. Insulation of an air space filled with a material of low bulk density (Kapok). If a guard-ring were used the lines would be straight.

air space of the required thickness. The graph also shows how reducing the radiation by reflecting faces now increases the insulation by only 15 per cent. Evidently the function of the filling material was merely to immobilize the enclosed air, preventing convection currents and making it effectively 'dead air'. It made no difference what filling material was used, whether cotton-wool, kapok, or even, in one experiment, extremely fine (000) steel wool, in each case the insulating value was the same, as long as the bulk density was low. This means that conduction of heat through the fibres of the filling material is negligible as long as they are fine enough and there are few enough of them. This was proved also by Larose (6) very beautifully by filling the space between the plates of a thermal-insulation apparatus with a wool double-pile fabric and altering its thickness by compression (Fig. 13). The thermal insulation was quite accurately proportional to thickness over the whole range, in which the bulk density was changed nearly four times. The slope of the curve is about 4·6 clo/in. This is also in accordance with earlier work by Speakman and Chamberlin (7).

The independence of the amount of fibres used as filler of the space is, of course, only over the range of low bulk densities. There seems to be an optimum density of about 4 lb./cu. ft. With lower bulk densities, the air is not completely immobilized. The same independence of the actual filling material has been shown with

E

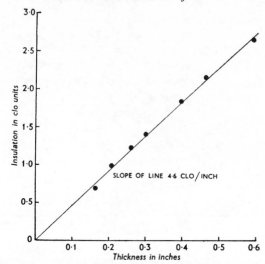

FIG. 13. Insulation and thickness of a wool double-pile fabric when the thickness was changed by compression. (From Larose. N.R.C. Canada, 1943).

low density building materials and all these fit closely to a universal curve of conductivity versus density. This means simply that the insulation is due to the 'dead air' enclosed and the solid material serves merely to immobilize it.

The value of 4·7 clo/in., or a conductivity of 0·00083 metric units, for dead air falls very far short of the insulation corresponding to the 'thermal conductivity of air' (0·000005) given in tables of physical constants, and this has caused some confusion of thought. It has not been realized that the values for the conductivity of gases are of no practical importance whatever, but are merely values deduced from difficult experiments of physicists in order to test molecular and kinetic theories. Whenever there is any temperature gradient, as there must be in a measurement of thermal conductivity, there will be movement of molecules from one region to another, carrying heat with them, and this type of transport is eliminated from the physicist's definition of 'conduction' of a gas. The value for conductivity in question was deduced by a series of experiments at different gas pressures, in which the inevitable convective loss was reduced at lower pressures, and the result was obtained by an extrapolation as to what the heat transfer would have been if the convective, and the radiative, losses were eliminated altogether (8).

The value 4·7 clo/in. seems to represent the insulation of still air where the convection currents have been reduced to an irreducible minimum. The only possible way of obtaining a better insulation than this is by using an evacuated space, as in the Dewar flask, or by reaching, by some other means, the condition where the mean free path of the air molecules is comparable with the dimensions of the spaces in which the air is restricted.* This is actually the case with the substance silica aerogel, in which the pores, containing the air at ordinary atmospheric pressure, are so small that the condition is realized, and the thermal insulation per inch of a space filled with this substance, a very breakable fine powder, actually is two or three times that of dead air. Since, however, aerogel is destroyed if it comes in contact with moisture, practical applications are limited to those where the material can be sealed hermetically from moisture.

* Kinetic theory predicts, correctly, that the thermal conductivity of a gas is independent of the density, until the mean free path becomes comparable with the dimensions of the enclosing space. Hence there is no gain when the space in a Dewar flask is pumped out, until the pressure becomes very low indeed.

REFERENCES

1. DuLong, P. L. and Petit, A. T. 1817. For an account of their researches on radiation and convection see Preston, T. 'The Theory of Heat,' 3rd Ed., pp. 504–512. Macmillan, London, 1919.
2. Burton, A. C. Standard Values for the Thermal Insulation of Ambient Air with Different Degrees of Air Movement, Nat. Res. Council, Canada, Associate Committee on Aviation Medical Research, Report No. 2753, 1944.
3. Winslow, C.-E. A., Gagge, A. P. and Herrington, L. P. The Influence of Air Movements upon Heat Losses from the Clothed Human Body. *Am. J. Physiol.*, **127**, 505, 1939.
4. Burton, A. C. and MacDougall, G. R. An Analysis of the Problem of Protection of the Aviator Against Cold and the Testing of the Insulating Power of Clothing. Nat. Res. Council, Canada, Report No. C, 2035, 1941.
5. Burton, A. C. Insulation by Reflection and the Development of a Reflecting Cloth. Nat. Res. Council, Canada, Report No. C, 2464, 1943.
6. Larose, P. Variation in Thermal Insulation of Double Pile Fabrics with Thickness. Nat. Res. Council, Canada, Report C, 2460, 1943.
7. Speakman, J. B. and Chamberlin, N. H. The Thermal Conductivity of Textile Materials and Fabrics. *J. Textile Inst.*, **21**, 29, 1930.
8. 'The Absolute Conductivity of Gases.' See Preston, T. 'The Theory of Heat.' 3rd. Ed. pp. 664–666. Macmillan, London, 1919.

CHAPTER 4

THE THERMAL INSULATION OF THE CLOTHING OR FUR, I_{cl}

The recognition of the rôle of 'dead air' in providing the best thermal insulation obtainable was of tremendous importance in the study and development of thermal insulation in clothing. When it was seen that the insulating properties of a fabric were due, not to the textile fibres directly, but to the dead air entrapped, much futile search for fabrics or fibres that would give better insulation per unit thickness, was eliminated. An 'aphorism' in this field is: *'The thermal insulation of clothing is proportional to the thickness of dead air enclosed.'*

Attention was thus diverted into much more profitable channels in studying the ways in which the thermal insulation of clothing could be improved. These are:

(*a*) In the design of garments the air entrapped should be immobilized as much as possible by preventing leaks of air which would allow air currents. The importance was realized of a windproof coverall, to prevent air currents within the clothing due to penetration of external air movement.

(*b*) The bulk density of materials used should be low enough so that the conduction by the textile itself was negligible.

(*c*) The maximum thickness of dead air entrapped should be maintained. This means choosing the materials and construction of fabrics so that the thickness is maintained under service conditions of compression by external forces (1, 2) and in conditions of moisture. The axiom shows that there is no intrinsic difference in thermal insulation which would allow us to choose, say, between wool and cotton except on the basis of this maintenance of thickness in all conditions.

A very great practical gain was achieved in some cases, particularly in gloves, by changing the conventional design to one which ensured the maintenance of thickness (3). The hand is, the great majority of the time, held with the fingers flexed, yet conventional gloves are tailored to fit the extended fingers, and when these are flexed, the thermal insulation at the joints is lost because it becomes thinner. The use of curved 'fourchettes' at once gave a glove

FIG. 14. The type E flying gloves of the R.C.A.F. in which by using 'fourchettes' which are curved to the usual position of the finger, thickness of insulation is preserved and freedom of movement improved.

warmer in use, and, in addition, giving greater freedom from restraint for the fingers (Fig. 14). The same principle applies to the design at knees and elbows of a garment for one who will usually be seated, as a pilot (4).

Again it was realized that the compressibility of pile fabrics could be decreased, without loss of flexibility, if a well-known engineering principle was applied. This is that the bending of a beam under load is proportional to a power of its length, the square. Thus if the 'ground fabric' of a pile is placed in the centre, with pile fibres extending on both sides, forming a double-faced pile fabric, the compressibility is very much decreased over a single-faced pile of the same total thickness, while flexibility is not lost. For details of this and of many other improvements in functional clothing which resulted from the fundamental principles outlined, reference should be made to the Report to the National Research Council of Canada by Kitching and Pagé (5), and to Part 11 of the book edited by Newburgh (6).

Regional Thermal Insulation—The Effect of Curvature

The use of a single value for the thermal insulation of the air, and of the whole clothing, of course, suffers from all the disadvantages and inadequacies intrinsic in any such average value. The necessity for freedom of movement of the limbs and other practical considerations, mean that the thermal insulation of clothing is very different in different parts of the body, and the face usually has to be left without any protection other than the insulation of the air. There is also a fundamental reason why the insulation that it is theoretically possible to provide for different parts of the body, is very different. This is the effect of curvature of a surface on its heat loss, first pointed out in connexion with the insulation of clothing by Van Dilla (7) and applied to the insulation of the air (8).

In the simplified equations of heat flow already used, it is assumed that the heat flows in parallel lines between two plane surfaces, so that the cross-sectional area through which it is flowing is invariable. In the case of heat flowing from a cylindrical or spherical surface, however, this is not the case. The area of cross-section of the flow is increasing as the distance from the surface increases. The modified law of heat flow is:

$$H = \frac{T - T_A}{f \cdot I} \quad \dots\dots\dots\dots\dots\dots\dots\dots\dots\dots\dots\dots1$$

i.e. the effective thermal insulation of the medium through which the heat is flowing is not the insulation I that applies to rectilinear flow, but is modified by a curvature factor f, which is less than unity. For a cylindrical surface,

$$f = ln\ r\ \left(1 + \frac{x}{r}\right) \quad \dots\dots\dots\dots\dots\dots\dots\dots\dots2$$

for a sphere

$$f = \frac{1}{1 + \dfrac{x}{r}} \quad \dots\dots\dots\dots\dots\dots\dots\dots\dots\dots\dots3$$

where r is the radius of curvature, and x the thickness of the insulation material added. For example, in a finger the radius of curvature r is about $\frac{3}{8}$ in. and a glove might have a thickness x of the same amount.

Van Dilla calculated that if x/r was 2 (a glove $\frac{3}{4}$ in. thick on a finger $\frac{3}{8}$ in. radius of curvature) the factor f would be 0·55 (for a sphere, 0·33) and the effectiveness of insulation correspondingly lowered. This, however, is not the full story of the effect of curvature. The increased area for radiation and convection into the air provided by the outside of the insulation layer means that the insulation of the air, if heat loss is reckoned per unit of the body, must also be modified. The result is that the total insulation, of clothing plus air, is greatly reduced in a cylinder of small diameter. This rather complicated matter is best understood from a graph of insulation against thickness (Fig. 15).

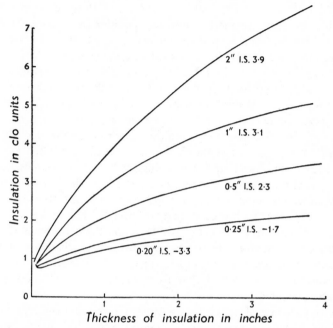

Fig. 15. Total insulation, including I_A, of perfect insulation on cylinders of small diameter. Showing how insulation is decreased on cylinders of small radius of curvature, even to the paradoxical effect, on very small cylinders, that total insulation may be decreased by the addition of an insulating layer. (From Burton, based on work of Van Dilla.) I.S.—initial slope in clo units per inch.

The decreasing slope as the thickness of insulation is increased is primarily due to the increasing cross-sectional area through which heat may flow. The initial slope of the graph represents the maximum clo/in. obtainable in the first thin layer of insulation. Even this is greatly reduced over the insulating power for recti-linear flow, and as the curves show, with cylinders or spheres of very small diameter, adding a layer of insulation, even of the best obtainable (4·7 clo/in.), may even increase the heat loss. This paradoxical result is because the area losing heat to the air is increased and this factor outweighs the small increase in the thermal insulation gained. This is familiar to physicists, who know that 'lagging' a pipe of diameter less than ¼ in. results in the opposite effect of what would be desired. Many engineers seem blissfully unaware of this. Assuming the ideal insulation (4·7 clo/in.) and still air, the maximum effective insulating power on a cylinder, that of the first layer of insulation, is reduced to 3·9 clo/in. on a cylinder of diameter 2 in. to 3·1 clo/in. on 1 in. diameter, 2·3 clo/in. on ½ in. diameter, and the paradoxical negative value (− 1·7 clo/in.) on a ¼ in. diameter cylinder. These values are the maximum, and in addition the insulating power of the succeeding layers of insulation falls off as the total diameter increases. For this reason, the insulation that can be provided for the fingers, or even for the whole hand by a mitt, is severely limited (9). While the factor of curvature is so important here, it is of little importance for the whole body, and the error in taking no account of curvature and using the standard curve already given for I_A will not amount to more than 0·1 clo/units. It has been calculated that a single cylinder representing an average for the whole body, would have a diameter of about 7 in., in which case the curvature effect is small. This means, however, that in 'models' designed as instruments to measure the effect of wind, etc. on the heat losses of the body, or to measure insulation of fabrics, the diameter, if the models are cylindrical, should be at least 7 in. Some erroneous conclusions have been drawn from experiments on models of smaller diameter.

The paradoxical effect that adding a layer of insulation to a cylinder of small diameter may increase the heat loss, may actually be realized in the case of thin gloves on the fingers. In still air, it can be calculated that even with the best insulation (4·7 clo/in.) the paradoxical effect on a finger would persist until a thickness of glove of ¼ in. was reached, after which there would be some gain in insulation. The coldness of thin gloves has often been noted, but has usually been attributed to restriction of the circulation by tightness. In high air movements, however, the paradoxical effect will not exist.

Thermal Insulation of Actual Clothing

It is interesting to see how closely we have been able to approach the theoretical ideal value of 4·7 clo/in. in clothing specially designed for thermal protection. It seems to be agreed that 4 clo/in. is approximately the value achieved in practice, when the clothing of the whole body is considered. There are two factors responsible for the failure to achieve the ideal. One is the effect of curvature in decreasing the effective insulating power in the case of the extremities, the other is the presence of air spaces between layers of clothing. If the air is not completely immobilized in such spaces, as Fig. 11 shows, the maximum insulation obtainable, whatever their thickness, will be about 0·5 clo units. This is about the value found experimentally to be added when a loose coverall was put over clothing (10). Siple and Cochran (11) developed a method of estimating the total thermal insulation of an 'assembly' of garments by measurements of circumference of limbs of a man wearing successive layers of clothing, from which the total thickness of 'dead air' enclosed could be estimated. Using the value of 4·0 clo/in. they showed that good agreement could be obtained, with direct 'clo determination'.

The practical value of 4 clo/in. for clothing insulation is of importance chiefly in that it tells us that, unless we used some material not utilizing the properties of dead air, we could not hope to achieve more than an increase of 16 per cent in the quality of clothing insulation. Secondly, this value shows us the practical limitation on the maximum clothing insulation obtainable. It is found that $1\frac{1}{2}$ in. of thickness of clothing would give a very bulky garment, and thus to provide more than 6 clo units, without seriously impairing the mobility of a man, seems to be almost impossible.

Insulation of the Fur of Animals

An interesting verification of the 'thickness law' of thermal insulation is provided by the work of Scholander *et al.* (12), who measured the thickness and thermal insulation, on a standard type of apparatus, of the fur of many tropical, temperate zone and Arctic animals (Fig. 16). The relation to thickness is approximately

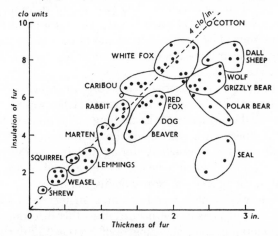

Fig. 16. Insulation of fur of animals, measured on a standard apparatus. (Redrawn from Scholander *et al. Biol. Bull.*, **99**, 225, 1950)

linear and the slope is about 3·7 clo/in., which is not too far from the ideal value for the air, and for our best fabrics. In the case of the seal the density of the fur is probably so great that conduction by the fibres is significant in lowering the insulating power, but the fur is very efficient in maintaining insulation when the animal is in the water.

Effect of Wind and Movement on Clothing Insulation

From the principle that the insulation of clothing is due to the 'dead air' entrapped, it is obvious that if the wind penetrates within the clothing and the air is no longer 'dead', a great deal of the insulation will be lost. Detailed results for different U.S. army Arctic uniforms are to be found in a report by Breckenridge and Woodcock (13). The thermal insulation was measured on an 'artificial man'. The effects of wind may be considerable, as for instance in a 'standard uniform', the insulation of which was 3·5 clo units in rather still air (2 miles/hr) but fell to 2·0 clo in a wind of 24 miles/hr. The magnitude of the effect depends on the wind resistance of the outer covering and on the efficacy of seals at the neck, wrists and ankles. Larose (14) made a valuable contribution when he related the fall of insulation of a fabric with external wind to the 'air permeability' of the covering fabric. Before this work we had no idea what permeability figure represented satisfactory 'wind-proofness'. Without any covering fabric, permeability infinite, the loss of insulation for a wind of 16 miles/hr was 1·0 clo, with a coverall of permeability 100 (cu. ft./sq. ft./min./pressure difference of $\frac{1}{2}$ in. H_2O) the loss was 0·7 clo, for a permeability of 10, 0·3 clo units. The exact results would of course depend on the insulating fabric underneath, the direction of incidence of the wind on the surface, and so on.

Even if there is no external wind, the 'internal wind' in the clothing caused by the bodily movements of the wearer can very greatly lower the insulation of the clothing. Thus it has been shown (15) that a typical U.S. army cold climate uniform possesses only about half as much insulation during walking as during quiet standing. This is not at all to be considered a disadvantage. Rather it is most advantageous, since the bodily movement connotes an increased heat production and requires an increased heat loss. Automatic adjustment in the insulation in this direction is to be sought in clothing, since the dangers of overheating during exercise in the cold and consequent loss of the insulation, because of dampness when the man is still, are recognized as the greatest problem of protection in the Arctic. Probably, in existing clothing, the automatic decrease in insulation with movement, though in the right direction, is not as great as would be desired, for while the heat production may increase to three times the resting value, the insulation of the clothing decreases to half. The eminently successful design of Eskimo winter clothing is superior perhaps because of this fact of automatic decrease of insulation when there is bodily movement. We would conclude that every effort should be made to prevent external wind from penetrating into the clothing, but, in contrast, we would encourage the production of an 'internal wind' in the clothing.

Moisture and Clothing Insulation

Practical experience suggests to all who have been interested and given thought to the matter, that when clothing is damp, it loses a considerable part of its power to insulate the body against cold. Yet many laboratory studies of the thermal insulation of fabrics have failed to show significant effects of increased humidity in clothing. The effect on heat transfer in the steady state seems to be significant only when actual liquid water has accumulated, and then only in proportion to the displacement of the 'dead air' in the fabric by this water. Of course in actual clothing there may be a change of thickness of the materials under the loads present when worn (2), and (16), if the resistance to compression of the fabric is decreased when damp. This may well explain the generally accepted superiority of wool over cotton for warmth in damp conditions, for wool certainly retains its thickness under load better than cotton when both are damp (17). However, there must obviously be more than this to explain the discrepancy between laboratory and field experience.

The nature of the difference appears to be emerging. It is that *in the wearing of clothing in practice, the steady state of temperature and humidity is never realized.* Clothing is always in the process of drying out or taking up more moisture. In the process of 'regain' of moisture, as is well known, textile fabrics take up or give up large quantities of water, to different degrees for different fibres; and in this process there is involved a relatively large amount of latent heat, plus a smaller heat of absorption specific to the particular fibre. The heat lost by the skin when the sweat and insensible water were evaporated there, is given back to the clothing if it is absorbed by the fibres. We can then see if in actual practice we have a 'non-steady' or changing state of temperature and humidity in the clothing, how the heat exchanges could be greatly affected by humidity, while in the measurements of thermal insulation, made deliberately in steady state conditions, these effects of humidity would be absent.

Evidence that the steady state is seldom reached in clothing emerged late in the last war in experiments by Kitching *et al.* (5), where the temperature and the relative humidity between different layers of clothing were measured. The latter measurement was made by removing small pieces of the fabrics at different times, and determining their water content (damp vs. dry weight). Calibration experiments on the damp weight of materials in equilibrium with different relative humidities enabled the relative humidity in the clothing to be estimated. Even where subjects sat quietly in the cold chamber in constant conditions for hours the temperatures and humidities were found still to be changing and very far from the steady state. In the ordinary use of protective clothing, where even if environmental conditions were constant for long periods, the metabolic heat production would be variable, we can safely assume that we never have to deal with thermal and vapour steady states.

The magnitude of the heat exchange involved in regain by fabrics is illustrated by calculations made by Cassie (18). Taking a man's wool suit of $1\frac{1}{4}$ kilograms weight, when the wearer moves from indoor conditions of 18°C (64·4°F) and 45 per cent R.H. to outdoors at 5°C (41°F) and 95 per cent R.H., 150 kilocalories of heat will be liberated as a result of the absorption of water by the suit (equal to the metabolic heat of $1\frac{1}{2}$ hours). The actual specific heat of the suit would contribute only 6 kilocalories. Thus although the thermal capacity of a suit of clothes, in the usual sense, is very small compared to that of the body wearing it (taking the specific heat of clothing as 0·4 cal/g/°C, a 12-lb. Arctic uniform would have a thermal capacity only 1/30 of that of the 180 lb. man wearing it), the clothing has a pseudo-thermal capacity which may be very large (up to that of the man himself). This is because any change of temperature of the textile fibres produces a change in relative humidity and therefore in regain, and this involves a large latent heat of evaporation.

Great progress is being made at the present time by Woodcock *et al.* (19) in the theory and experimental verification of these remarkable effects of the humidity in clothing on heat exchanges. They show that they depend on whether the fabric of clothing is 'bibulous', i.e. capable of transferring the absorbed water from one part to another by 'wicking', or 'non-bibulous'. As a consequence of these effects, the heat loss incurred in the evaporation of moisture from the body or from the clothing can be either considerably greater or considerably less than that calculated from the latent heat of the water evaporated. The theory is little understood by others and the work is in its exploratory stages, so it is not appropriate to give details here. The answer to very important practical problems of protecting man in the cold may emerge. Some of the conclusions are that for perfectly wicking materials, ('bibulous'), the rate of heat loss is constant during the drying period and then drops suddenly to the normal value for dry insulation when all moisture is evaporated. On the other hand, for perfectly non-wicking, 'non-bibulous' materials, the rate of heat loss decreases steadily during the drying period. Because of this, the additional heat loss is much less than the latent heat of the water evaporated, as low as one half. However, drying takes longer to complete than with a wicking material. Experimental results have agreed with these predictions.

Exactly analogous problems of the effects of moisture on the passage of heat through building materials used for insulation have been studied by Pfalzner (20). There would be great profit in conferences between the two groups of workers.

The subject is a complicated one, but it seems worth while to explain here the origin of the complications. Within the clothing two gradients exist at the same time, the thermal gradient from skin temperature to the environmental temperature, and the vapour pressure gradient from the skin to the environment. Let us take the simplest case possible, where these gradients are both linear. This would only be so where both the thermal insulation and the vapour resistance of the clothing were uniform throughout. Actual cases would have gradients that showed changes of slope according to the insulation and vapour resistance of the various fabrics.

The simple case is illustrated by Fig. 17 (*a*) and (*b*). In each case a reasonable skin temperature and humidity is assumed. (*a*) and (*b*) differ in that the thermal gradient in (*b*) is much steeper because of the lower environmental temperature. The regain of textile fibres depends not upon the vapour pressure, but upon the relative humidity of the place where the fibres may be, and then is almost independent of the temperature at that place (18). From the graphs in Fig. 17 we can read off the temperature and vapour pressure for different distances from the skin. The temperature gives the saturated vapour pressure from standard tables. This falls very rapidly with lowered temperature and within 0·5 mm. Hg is given, between 40°C and 0°C (104°F − 32°F) by the equation,

$$\text{Log}_{10}(\text{Sat. V.P.}) = 9 \cdot 1337 - \frac{2312}{T} \quad \dots\dots\dots\dots\dots 4$$

where T is the absolute temperature (273°C).

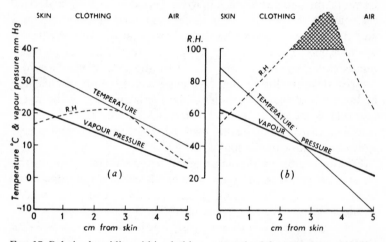

Fig. 17. Relative humidity within clothing as a result of the temperature and the vapour pressure gradients. In the hatched area water would continuously condense.

The relative humidity is given by the ratio of the vapour pressure, at the point concerned, to this saturated vapour pressure. The broken lines in Fig. 17 show that, astonishingly to some, the relative humidity rises as we pass out from the skin and reaches a maximum at a certain distance from the skin. Where the temperature gradient is very much less, the relative humidity may decrease to a minimum instead. While in Fig. 17(a) this maximum would be 62 per cent, if the environment were colder as in (b) the maximum will exceed 100 per cent relative humidity. Though the vapour pressure has decreased as we leave the skin, the decrease is not enough to keep pace with the rapidly falling capacity of the air to hold the moisture as the temperature falls. Where the relative humidity exceeds 100 per cent, there will be continuous condensation of moisture, from the skin, in those particular layers of the clothing. This curious phenomenon of wetting out, or of a 'frost-line' where the temperature is below freezing, at a certain level in Arctic clothing or in the insulation of a sleeping bag, has been often noted by those with experience in living in the cold. Unless the frost, or moisture, can be eliminated by shaking or by drying, it will be cumulative, and the weight of clothing or a sleeping bag become very great in a few days.

We cannot therefore, even if we were dealing with the steady state, treat the thermal gradient and the vapour gradient separately in the study of clothing insulation. The vapour permeability of fabrics in practice profoundly affects their thermal insulation, and conversely, their thermal insulating power affects the transfer of vapour through them.

Vapour Permeability of Fabrics

As already mentioned in Chapter 3, this was first studied in the last war by Goodings and co-workers and later by Fourt and Harris. Much study of the relation of structure of fabrics as well as the nature of the fibres to the vapour permeability remains to be done, but certain rather unexpected conclusions have been reached. The most important of these is that in general there is no correlation of the vapour permeability of a fabric and its air porosity. Some very tightly woven fabrics, such as 'Grenfell cloth' or 'Byrd cloth', used in windproof coveralls and jackets, are found to have greater permeability to vapour, if the fabric be thin, than the much thicker fabrics of open weave, such as have been advertised as desirable for tropic wear.

Movement of air through fabrics, which is concerned in air porosity, is a process of 'filtration'. In contrast the movement of water vapour down a gradient of vapour pressure which is concerned in vapour permeability is by 'diffusion'. These two processes are quite distinct and the physical laws that describe them have a different dependence upon the radius of the pores through which both proceed. Pappenheimer (21) has shown, for the analogous processes of filtration and of diffusion through the pores in the wall of the blood capillary, that the ratio of the filtration constant to the diffusion constant depends upon the pore size. The smaller the pores, the greater will be the diffusion constant compared to the filtration constant. He has used this relation to deduce the size of the pores in the wall of the capillary. There seems no reason why the same type of analysis should not explain the rôle of the pore size and the distance between pores in the case of fabrics, in determining the relation of air porosity to vapour permeability. The rôle of the pore size and the distance apart of the pores has still to be experimentally elucidated.

Where sweating is apt to occur, as on the soles of the feet of a man under emotional tension or any kind of discomfort, and certainly in battle, the provision of adequate vapour permeability is necessary. The astonishing result of an extensive field trial on the comfort factors in boots (22) revealed that the vapour permeability of the leather or plastic *soles* of the boots demonstrably affected the comfort of soldiers, even in cold conditions.

Vapour Resistance of Clothing

Very little data are available yet on the total resistance to the passage of water vapour of clothing assemblies. Enough has been done, however, to show that the choice of '1 cm. of dead air' as the unit of vapour resistance is indeed a happy one. It appears that the vapour resistance of the normal indoor clothing, 1 clo in thermal insulation, is probably of the order of magnitude of '1 cm. of dead air', though we would expect the variability from one assembly to another to be very great. Also the vapour resistance of the air, R_A, appears to be about 0·8 cm. of dead air for low air movements, decreasing to about 0·2 cm. for high air movements. If further measurements confirm this, the analogy with clo units of thermal insulation will be quite complete. It can be calculated that for the non-sweating man at rest, the total vapour resistance could be as high as 17 cm. before there was discomfort. With a man in physical exercise or in hot surroundings, the total resistance must, of course, be much less if he is to achieve a steady state.

The Physiological Efficiency of Evaporation

If the moisture that is evaporated at the skin condenses in the clothing at a certain level, 'wicks' to the surface of the clothing, and re-evaporates there, we cannot consider that all of the heat of vaporization of the water that has been evaporated has come from the body. The condensation of the vapour in the clothing raises the temperature there, so that the gradient of temperature from skin to clothing is less than it would otherwise be and less non-evaporative heat is lost by the body. In contrast, at the surface of the clothing the evaporation once more takes place, and this lowers the temperature of the clothing so that less non-evaporative heat passes from surface of clothing to the environment than if the clothing were dry. Thus we may consider that part of the heat of evaporation came, not from the body but from the environment, in that less heat passing from clothing to environment is equivalent to the normal heat loss plus an item of heat from environment to clothing. Examination of the equations of heat loss for this case, (23), shows that the portion of the heat of evaporation that can be considered to be effective in cooling the body is given by the factor f_E.

$$f_E = \frac{\text{Total dry insulation (of dry clothing + air)}}{\text{Total insulation (of all clothing + air)}}$$

Or in symbols, the physiological efficiency of evaporation:

$$f_E = \frac{I_{cl.D} + I_A}{I_{cl.D} + I_{cl.W} + I_A} \quad \dots\dots\dots\dots\dots 5$$

where I_{cl} is the thermal insulation of the dry clothing between the skin and the point of condensation, and I_{clW} of the wet clothing.

This is of very great importance to the heat exchanges of man in hot environments, since here the evaporative loss of heat becomes the major item of the total heat loss. It is of importance to understand that the effect exists in estimating heat loss in the cold also, although, as has been emphasized, this is a 'steady-state' effect and the 'non-steady state' effects of moisture in clothing are of much greater practical importance. It is interesting that, theoretically at least, it is possible that the condensation of moisture in clothing, which then 'wicks' out to the surface and is re-evaporated there, could conceivably increase the protection of the wearer against cold. If the loss in the physiological efficiency of evaporation outweighed the loss in total insulation of the clothing, the total heat loss might actually be decreased. The principle of the division of heat items between the environment and the body, and the concept of 'efficiency' in this sense, is used also in a later chapter in connection with the utilization of solar radiation by the body.

REFERENCES

1. KITCHING, J. A. Compression Loads to which Material Forming Parts of Articles of Clothing are Subjected. Nat. Res. Council, Canada, Report No. C 2294, 1942.
2. KITCHING, J. A. Report on Comfort and Service Stresses in Flying Suits. Nat. Res. Council, Canada, Report No. C. 2295, 1942.

F

3. KITCHING, J. A., BENTLEY, A. N. and PAGÉ, E. New R.C.A.F. Flying Gloves. Nat. Res. Council, Canada, Report No. C. 2216, 1942.

4. WEBB, P. W. The Type E. Flying-Suit, Nat. Res. Council, Canada, Report No. C. 2497, 1943.

5. KITCHING, J. A. and PAGÉ, E. Review of the Work of the Sub-Committee on Protective Clothing. 1942–1945. Nat. Res. Council, Canada, Report No. C. 3039, 1945.

6. NEWBURGH, L. H. 'Physiology of Heat Regulation.' Saunders, Philadelphia, 1949.

7. VAN DILLA, M. Effect of Curvature on Heat Loss, Climatic Research Laboratories, U.S. Quartermaster Corps. Report No. 76A, 1944.

8. BURTON, A. C. The Effect of Curvature on the Heat Loss from the Body, with Special Reference to Handgear. Nat. Res. Council, Canada, Report No. C. 2725, 1944.

9. VAN DILLA, M., DAY, R. and SIPLE, P. A. In 'Physiology of Heat Regulation.' Newburgh, pp. 374–386. Saunders, Philadelphia, 1949.

10. BURTON, A. C. and COLES, B. C. Note on the Nature of the Thermal Insulation added to a Flying-Suit when a Coverall is Worn. Nat. Res. Council, Canada, Report No. C. 2297.

11. SIPLE, P. A. and COCHRAN, M. I. See Conference of Principles of Environmental Stress on Soldiers, p. 14–17. Environmental Protection Section. Office of Quartermaster General, 25th August, 1944 (Re-issued February, 1947).

12. SCHOLANDER, P. F., WALTERS, V., HOCK, R. and IRVING, LAURENCE. Body Insulation of some Arctic and Tropical Mammals and Birds. *Biol. Bull.*, **99**, 225, 1950.

13. BRECKENRIDGE, J. R. and WOODCOCK, A. H. Effects of Wind on Insulation of Arctic Clothing. Environmental Protection Section, Quartermaster Climatic Research Laboratory, Report No. 164, 1950.

14. LAROSE, P. The Effect of Wind on the Thermal Resistance of Clothing with Special Reference to the Protection given by Coverall Fabrics of Various Permeabilities. *Canad. J. Res.*, **A 25**, 169, 1947.

15. BELDING, H. S. Protection Against Dry Cold. Office of the Quartermaster General Environmental Protection Section. Report No. 155, 1949.

16. KOLESAR, H. M., BURTON, A. C. and GOODINGS, A. C. The Effects of Repeated Compressions and Recovery, such as occur in Wear of a Garment, on the Thickness of a Double Pile Garment. Nat. Res. Council, Report No. C. 2526, 1943.

17. LAROSE, P. Change in Thickness of Double Pile Fabric with Moisture Content. Nat. Res. Council, Canada, Report No. C. 2654, 1944.

18. CASSIE, A. B. D. Physics and Textiles. *Reports on Progress in Physics*, **10**, 146, 1946.

19. WOODCOCK, A. H. and DEE, T. E., Jr. Effect of Moisture on the Transfer of Heat Through Insulating Materials. Office of the Quartermaster General, Environmental Protection Section No. 170, 1950.

20. PFALZNER, P. M. On the Flow of Gases and Water Vapour Through Wood. *Canad. J. Res.*, **A 28**, 389, 1950.

21. PAPPENHEIMER, J. R. Passage of Molecules through Capillary Walls. *Physiol. Rev.*, **33**, 387, 1953.

22. HITCHMAN, N. and BURTON, A. C. Unpublished Observations.

23. BURTON, A. C. An Analysis of the Physiological Effects of Clothing in Hot Environments. Nat. Res. Council, Canada, Report No. 186, 1944.

CHAPTER 5

THE THERMAL INSULATION OF THE TISSUES OF THE BODY

In the equation of the thermal steady state, the insulation of the tissues, I_T, denotes the over-all mean insulation, i.e. the total resistance to flow of heat from the 'core' of the body to the body surface. This over-all average is the resultant of the resistances of a number of pathways for heat flow, which may be considered as in parallel with each other, from the core to different regions of the body surface. Again, each of these pathways will consist of a number of elements in series with each other. For example, these are for the heat reaching the surface of the hand, the insulation to the flow of heat down the length of the arm, that from the arteries to the distributing subcutaneous vascular bed, and finally the resistance through the skin itself to the surface. Thus, although the over-all average insulation of the tissues is a convenient concept in studying the total thermal economy of the body, its dependence on the physiological factors can be very complicated. Neglect of the underlying complexity can lead to confusion, and experimental data seem to be contradictory, because the particular part of the thermal pathway with which the various experiments deal may be different.

Thermal Insulation of the Skin

The thermal insulation, or its inverse, the thermal conductivity, of the last barrier in the total thermal pathway, that is of the skin itself, has been the object of some study. However most of the research has been upon dead skin, and obviously the blood flow of the skin must play a large part in determining its effective conductivity in life. The blood flow of the skin shows an enormous physiological variability, from less than 1 ml/min/100 ml. of tissue to over a hundred times this value (1) and this may be expected to give a great variability to the thermal insulation of the skin.

Heat is carried to the surface through the skin both by the physical process of conduction from cell to cell, and by convection of heat by the blood flowing from warmer to cooler tissues. We might consider these two modes of transport either as being in parallel with each other, or as in series in the skin. The two theories would give different predictions as to the relation to be expected between the effective insulation and the amount of the blood flow. A strictly parallel system would give a linear relation between conductivity and flow, while a strictly series system would predict that the conductivity would increase linearly with flow at low flows and increase more slowly with greater flows. It is unlikely that present or future data can decide the point, especially as the actual arrangement must be a complicated combination of these two modes of arrangement, i.e. a 'compound' circuit arrangement.

The Thermal Conductivity of Dead Tissue

As long ago as 1892, Bordier (2) compared the thermal conductivities of several kinds of dead tissue by using slices of 1 mm. thickness in a standard physical apparatus. He compared their conductivities to that of the empty air space in the gap in the apparatus in which he placed them. His results were as in the table below.

RELATIVE THERMAL CONDUCTIVITIES OF TISSUES

Air gap	1·00	Muscle (longitudinal section)	2·48
Bone	4·45	Tendon	1·92
Muscle (perpendicular section)	2·83	Cartilage	1·85
Blood clot (after 24 hours)	2·71	Adipose tissue	1·38

Lefèvre (3) criticizes these values on the ground that the conductivity of tissues must be very much greater than that of air, evidently thinking of the physicist's value for the pure conductivity of air. If we assume the value of 4·6 clo/in. for the insulating power of the static air on the narrow air gap of Bordier's apparatus, we would have 3·3 clo/in. for fat. This value proves to be far greater than has since been found, so we must reluctantly conclude that Bordier's results cannot be of absolute value, though they do show that adipose tissue is about twice as good a thermal insulator as muscle.

Since all of the workers in this field use c.g.s. units of conductivity and we wish to relate the insulating power of the tissues to the clo units used in calculating the heat balance and compare with that of clothing, we must give the conversion factor. This is that *an insulating power of* 1 *clo/in. thickness corresponds to thermal conductivity of* 0·000393 *c.g.s. units.* Conversely, a thermal conductivity of 0·001 c.g.s. units would correspond to a thermal insulating

power of $\dfrac{0\cdot000393}{0\cdot00100} = 0\cdot393$ clo/in.

Lefèvre made some remarkable studies of the thermal conductivity of the skin and subcutaneous layers in man in water baths and found a mean value for the conductivity of the skin of 0·00060 c.g.s. units (0·65 clo/in.). In cold baths there was vasoconstriction, and the thermal conductivity of the skin fell to a minimum of 0·00047 c.g.s. units (0·84 clo/in.). These values, of course, were for live skin with its blood flow, and the insulating power of dead skin might be expected to be a little higher than the maximum found in cold baths.

For dead adipose tissue we have the observations of Scholander *et al.* (Chapter 4) on seal blubber, which is 2 to 3 in. thick and poorly vascularized. Its insulating value was about one-quarter that of cotton batting, i.e. about 1·25 clo/in. (k = 0·0003 c.g.s. units). In life the insulating power must be considerably less because of the blood flow. Recently there have been new measurements on dead tissue with particular interest in the relative insulating power of fat compared to muscle. Hardy and Soderstrom (4) found, for beef tissue, the values 0·00049 for fat (0·80 clo/in.) and practically the same value, 0·00047, for muscle (0·83 clo/in.) and concluded that the supposed extra insulation of fat people was not significant. However Hatfield and Pugh (5), using slices of human tissue from necropsy, found a much greater difference between fat (0·00049) and muscle (0·00092), indicating that fat (without blood flow) was twice as good an insulator. They noted, as also with beef tissue, that while the measurements on fat did not change with time, those for muscle increased markedly when it was kept for a day or two at room temperature. Recent determinations suggest that a rough rule would be that the thermal insulation of dead fat, skin and muscle are in the ratio 3 : 2 : 1.

Measurement of Thermal Conductivity of Skin in Vivo

A method has been developed (6), by which direct estimates of the effective thermal conductivity of human skin can be made by an apparatus which is essentially a 'surface thermo-stromuhr'. Since this has never been published, a description follows. Two flat resistance coils, the smaller a central disc and the larger a ring about it, with an insulating gap between, form the two arms of a Wheatstone's bridge. These coils are of wire which has a high temperature coefficient of resistance, while the other two coils of the bridge are of constant resistance. The battery supplying the bridge is arranged to drive current through the two 'skin' coils in

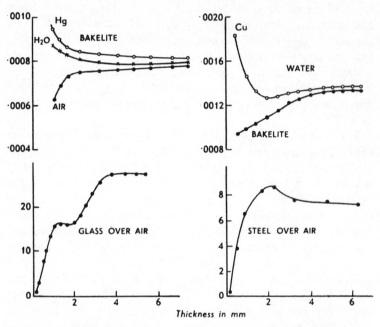

FIG. 18. Estimation of the depth to which the measurement of thermal conductivity is made by the apparatus. The results with the various materials over water, mercury or air, agree in suggesting 4 mm as this depth.

parallel. Since the central coil is of lower resistance, greater heat is generated since the current is greater in it, and as a result it will reach, in the thermal steady state, a higher temperature above the skin than the other, 'ring' coil. The difference of temperature between the two coils is registered directly by the deflection of the galvanometer of the bridge. The magnitude of the difference of temperature, for a constant current in the bridge, will be determined by the difficulty with which heat flows from the central to the ring coil, that is, it depends upon the effective thermal conductivity of the surface upon which the coils are in contact (not at all upon their temperature). The apparatus was calibrated by measurements on samples of metal, glass, rubber, etc., down to 'dead air' (cotton-wool) whose conductivity was known. The deflection was shown to be directly proportional to the thermal insulating power of these materials. Thereafter water ($k = 0 \cdot 0014$) and glycerine ($k = 0 \cdot 00068$) were used as convenient substances for calibration, as these values lie towards the two ends of the range found for human skin.

It is important, since the structure and vascularity of skin is so different at different depths, to know to what depth such an apparatus may be considered to measure conductivity. A direct demonstration of this was made by placing the apparatus on slabs of different materials with other materials beneath. The thickness of the top material was then progressively decreased, by milling off a layer at a time, and the thickness at which the deflection of the instrument began to be affected could be noted. Typical results are shown in Fig. 18. The complexity of some of the curves need not be discussed here, since all agree in suggesting that the thermal conductivity is increased down to a depth of about 4 mm. Of course, the sensitivity of the instrument to the conductivity of a given layer of the material must depend greatly on its depth, but the heat flow between the two coils evidently does not penetrate significantly below 4 mm. This is about equal to the size of the gap between the two coils.

To discover the relation between the thermal conductivity, so measured, and the blood flow of the skin, the fingers were utilized as test objects, since here the total flow, which can easily be measured in ml./min by venous-occlusion plethysmography (1), is all to the skin except for a very small fraction to bone. Simultaneous measurements of conductivity and flow in fingers, in a variety of different physiological conditions, gave the graphs shown in Fig. 19, hitherto unpublished. These are for three different subjects. The correlation is excellent and the relation appears to be linear. The total range of thermal conductivity of the skin, to a depth of 4 mm. is to about 160 per cent of the value for no flow. Pure conductivity, without convective transfer, is therefore always a considerable item in the total heat transfer of the skin. Differences between different subjects might be attributed to different contents of water and of fat.

With the same apparatus, measurements were made in the same subject for different areas of the skin with the subject cool and presumably moderately constricted, and also when he was warm

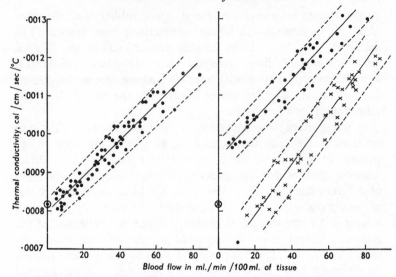

FIG. 19. The relation between effective thermal conductivity of the skin and blood flow in the fingers. The three curves are for different subjects. The points for zero flow, circled by a ring, were obtained by occlusion of the flow by a cuff on the proximal phalanx, while the other data for low flows were obtained by vasoconstriction to cold.

and the peripheral vessels were dilated. The results are given in the following table.

CHANGES IN SKIN TEMPERATURE AND EFFECTIVE THERMAL CONDUCTIVITY OF SKIN IN DIFFERENT
REGIONS OF THE BODY SURFACE

	Skin temp. °C (°F)				Thermal conductivity × 10⁻⁶ c.g.s. units		
Room temp. °C (°F)	25°C (77°F)	30°C (86°F)	Rise °C	°F	25°C (77°F)	30°C (86°F)	Rise
Forehead	34·6 (94·2)	35·5 (95·9)	0·9	(1·62)	1175	1245	70
Upper arm	33·6 (92·5)	34·6 (94·2)	1·0	(1·8)	880	1140	260
Lower arm	33·6 (92·5)	34·6 (94·2)	1·0	(1·8)	930	1110	180
Palm of hand	33·5 (92·3)	35·0 (95·0)	1·5	(2·7)	900	1110	210
Index finger	32·2 (90·0)	35·0 (95·0)	2·8	(5·0)	906	1120	214
Sternum	34·5 (99·1)	35·0 (95·0)	0·5	(0·9)	837	1085	248
Umbilicus	34·1 (93·4)	34·9 (94·8)	0·8	(1·44)	868	1030	162
Thigh	32·2 (90·0)	34·1 (93·4)	1·9	(3·4)	816	968	152
Calf	32·9 (91·2)	33·4 (92·1)	0·5	(0·9)	844	952	108
Dorsum of foot	30·4 (86·7)	32·9 (91·2)	2·5	(4·5)	930	992	62
Great toe	25·3 (77·5)	35·5 (92·3)	8·2	(14·8)	830	1045	215

These results illustrate how, while the changes in skin temperature between vasoconstriction and vasodilatation differ greatly in different parts of the body (0·5°C (0·9°F) rise for the sternum vs. 8·2°C (14·8°F) rise for the toe), the changes in skin thermal conductivity show much smaller differences. The rise for the sternum is just as great as for the toe.

There are also recent studies of the penetration of the increased temperature to depths below the skin when a small area is heated by radiation, and also observations of the periodic fluctuation of temperature below the skin when an applicator was applied which was periodically heated and cooled (U.S. Army reports). This last method is the one originally used by Ångstrom to measure the thermal conductivity of the earth's crust, utilizing the naturally occurring periodic change of temperature between day and night. The results are hard to reconcile with other data. The question of how much the application of the heat or cold elicits a reaction in the skin being measured, does not seem to be satisfactorily discussed. Reader (7) recently measured the temperature at different depths below the skin, to 40 mm. depth, by needle thermocouples, while a cold applicator, cooling temperatures from 27°C (80·6°F) to 28°C (82·4°F), was applied to the skin of the back. The applicator had a 'guard ring' and the rate of heat flow from the skin to the applicator could be measured. Thus he calculates the conductivity of different layers of tissue. In general the apparent conductivity increased as between the layers 0–10 mm. deep, 10–20 mm. deep, and 20–30 mm. deep. His mean value of conductivity for skin, presumably with minimal blood flow because of the cooling, was 0·00075 c.g.s. units and 0·00127 for muscle.

One serious objection to the interpretation of data such as these in terms of conductivity, which applies to all experiments where a different temperature, hotter or colder, than the normal is applied to an area of skin, is that in these circumstances the flow of heat from deeper layers to that area must be far from rectilinear. When the area is cooled, heat must flow in to it from surrounding areas outside the ideal cylinder of tissue imagined for rectilinear flow. Thus the figures for the conductivity of deeper areas must be greatly in excess of the true values, because the cross-sectional area through which the heat is flowing, at depth, is much greater than that of the applicator. The 'guard-ring' on the applicator improves the method, but what is needed is a 'guard-ring' down to the deep tissues. When the normal gradients in the tissue are not seriously disturbed, the lines of flow of heat to any given area of skin are much more likely to be normal to the surface.

Insulation of Subcutaneous Tissues

Lefèvre (3) made some remarkable experiments on the changes in the heat transfer coefficient not only in the skin but also in subcutaneous tissues. With his subjects in a water bath he measured the temperature of the water in the bath, of the skin surface, and of the subcutaneous layers, the latter by the insertion of needle thermocouples. He also measured the heat traversing the tissue and reaching the water, by calorimetric methods. Thus he could calculate the effective thermal conductivities from the skin to the water, from the epidermis, and in the subcutaneous tissues (he assumed the total temperature gradient for the latter to be from 37°C (98·6°F) to the subcutaneous temperature). His results for experiments made in baths from 5°C to 30°C (41°F to 86°F) were:

Temperature of bath, °C	Effective Conductance* skin to bath	Effective Conductance of skin	Effective Conductance of subcutaneous tissue
5	·0016	·00047	·00120
12	·0015	·00058	·00070
18	·0015	·00066	·00045
24	·0015	·00075	·00031
30	·0016	·00083	·00021

Thus while the thermal insulation of the skin increased about twice as the temperature of the bath fell from 30° to 5°C (86°F to 41°F), as one might expect because of vasoconstriction, the thermal insulation from the core of the body up to the skin apparently *decreased* at the same time by a factor of six times. This puzzling result has not been confirmed. At any rate we know that such a decrease in 'deep tissue insulation', if it occurs in the cold, does not prevent the total insulation of the tissues from increasing (see next section). Lefèvre's results, which are perhaps open to question as to their application to the whole body from experiments in one locality, do emphasize however that the changes in insulation at different depths in the skin can be very complicated. Remarkable anomalous temperature gradients in the skin have been measured by Bazett (8), particularly when the steady state has not been reached. Possibly the error in interpretation of the results of Lefèvre is to assume that the data represent a steady state at all. If vasoconstriction in the limbs is accompanied by a shift of the venous return from superficial to deep venous channels, as Bazett suggests, the results of measurements of the effective conductivity

* Conductance is in the same units as conductivity except that the thickness factor is omitted.

of skin by apparatus such as described in a previous paragraph may be very greatly affected by the depth to which the measurement is made.

We may conclude from all these studies that the regulation of the effective thermal conductivity of the last part of the thermal pathway, the skin itself, is not the important mechanism in controlling the total insulation of the tissues from core to surface. For, as shown in the next paragraphs, the change in the total insulation between vasoconstriction and vasodilation is of much greater magnitude than is found for the skin alone. The regulation of thermal insulation down the length of the limbs is much more important than is the regulation of the gradient from deep tissues up to the skin at the end of the thermal pathway.

Total Insulation to Different Skin Areas

The insulation of the tissues used in the general equation of heat exchange for the whole body, could hardly be deduced from the values for local thermal conductivity. The whole thermal pathway from the core of the body to a particular area of the body surface is here involved, and for the figure for the whole body, a mean value for all such areas of surface. That great differences in thermal insulation of the whole thermal pathway to different skin areas exist is shown by examination of the 'thermal circulation index' (9). This index is based on the equations of heat flow up to and away from a given area of body surface. Let the temperature of the area of the skin be T_S, of the 'core' be T_c, and of the air T_A. Then we have two equations:

$$(a) H = \frac{T_c - T_S}{I_T} \text{ where } I_T \text{ is the insulation of the tissues}$$

for that area of skin, for the flow up to that area, and

$$(b) H^1 = \frac{T_S - T_A}{I_A} \text{ for the heat flow away from that area,}$$

into the environment, where I_A is the insulation of the air for that same area of skin. This will be different, for the areas where the curvature is pronounced, from the general value of I_A for the whole body.

If there is no sweating, we can assume, as a first approximation, that the heat loss from the skin by evaporation bears a constant ratio to the total heat loss from the skin, i.e. $H^1 = 1 \cdot 21 H$ (see reference 9).

We then can eliminate H and H^1 and have

$$\frac{I_T}{I_A} = \frac{T_c - T_S}{T_S - T_A} \times 1{\cdot}21 = 1{\cdot}21 \times \frac{\text{Internal drop of temperature}}{\text{External drop of temperature}}$$

The ratio of the external drop of temperature to the internal drop of temperature has been called the 'thermal circulation index'.

A simple nomogram (Fig. 20) can be used to read off the thermal

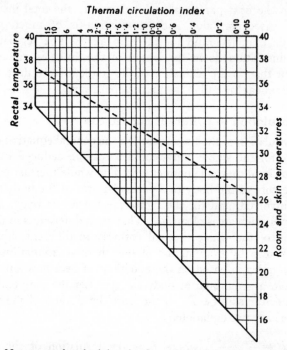

Fig. 20. Nomogram for obtaining the thermal circulation index from the skin, rectal and room temperatures. Place a tight thread or ruler connecting the rectal temperature (left-hand vertical scale) with the room temperature (right-hand vertical scale), as indicated by the broken line. Proceed horizontally from the skin temperature (right-hand vertical scale) to the intersection with the connecting line. The position of the intersection as to the vertical lines gives the thermal circulation index. A doubling of the index indicates an increase of circulation to that part of at least twice.

circulation index from the skin temperature, room temperature and rectal temperature. Measurement of the skin temperature on different areas in a subject in a steady state will then allow us to calculate the ratio I_T/I_A for the different regions, and see how it changes with changes in physiological conditions, as in general vasoconstriction and general vasodilation. If the air movement is constant and therefore I_A has not changed, the change in the thermal circulation index will indicate the extent of physiological changes of the insulation of the tissues I_T. In fact, if we can assume a value for I_A for each area (the curvature effect will have to be taken into account) we can deduce a value in clo units for insulation of the tissues up to the different regions of skin. This has been done in the following table, based on the data obtained some years ago (9) before there was any knowledge of the values of I_A.

SKIN TEMPERATURES AND THERMAL CIRCULATION INDICES OF DIFFERENT AREAS ON SUBJECT
AFTER LYING 20 MINUTES AT ROOM TEMPERATURE OF 22·8°C (73°F).
RECTAL TEMPERATURE 37·25°C (99°F).

Location	Skin. temp. °C (°F)	External drop of temp. °C (°F)	Internal drop of temp. °C (°F)	Thermal circulation Index	Calculated insulation of tissues I_A clo
Forehead	33·40 (92·1)	10·60 (19·1)	3·85 (6·9)	2·75	0·26
Clavicle	33·60 (92·5)	10·80 (19·44)	3·65 (6·6)	2·96	0·24
1 in. above umbilicus	34·20 (93·56)	11·40 (20·5)	3·05 (5·5)	3·75	0·19
Lumbar region	33·30 (91·9)	10·50 (19·0)	3·95 (7·1)	2·67	0·26
Arm, biceps	32·85 (91·1)	10·05 (18·1)	4·40 (7·9)	2·28	0·31
Knee cap	32·35 (90·2)	9·55 (17·2)	4·90 (8·8)	1·95	0·36
Calf of leg	32·20 (89·96)	9·40 (16·9)	5·05 (9·1)	1·86	0·38
Sole of foot	30·20 (86·36)	7·40 (13·3)	7·05 (12·7)	1·05	0·66
Big toe	30·95 (87·7)	8·15 (14·7)	6·30 (11·3)	1·29	0·39

Thus it appears that the insulation of the thermal pathway from the core of the body to the big toe is about twice that of the thermal pathway to the surface of the abdomen. Further data obtained with a range of environmental temperature showed a 20 per cent increase in thermal circulation index for the skin of the trunk and a 30 per cent increase for the leg in the warmer conditions as compared to the colder. Under complete general anaesthesia the index for the toes increased as much as ten times; the toe temperature rose to 33°C (91·4°F). It is apparent how much more variable is the insulation of the tissues to the extremities than to the rest of the body, and the point has often been made that the extremities are the regions where the 'physical regulation' of body temperature is mostly accomplished.

Although the idea is familiar that the decrease in the internal gradient, as by a rise of skin temperature, indicates a decrease in the tissue insulation and vasodilation, this has been misinterpreted in some cases. The amount of heat, H, flowing down that gradient has also to be taken into account. Thus in exercise, where the metabolic rate is increased, the heat flow must be greatly increased. In this case, even though the skin temperature may be prevented from rising, due to cooling by the large amounts of evaporation of sweat, the insulation of the tissues can be shown to have decreased greatly, and there is good evidence of vasodilation. Two or three times as much heat is flowing down the same temperature gradient as before. Erroneously the lack of rise of skin temperature in exercise, compared to exposure to hot environments, has been taken to indicate that vasodilation and decrease of tissue insulation is not of the same importance in temperature regulation in exercise as in thermal stress (10). Use of the insulation of the tissues shows this is mistaken. Actually, the thermal insulation of the tissues decreases to about the same extent in exercise as in hot environments.

General Insulation of the Tissues

The changes in the mean conductance of the tissues for the whole body, excluding the head only, have been measured in subjects in well-stirred water baths fitted with apparatus to record the heat given by the body to the bath (11). Dividing the internal gradient by the measured heat gives the insulation. The results were expressed in terms of the heat transfer coefficient or effective conductance down the internal gradient (cal/sq. m/hr/°C differences of temperature from core to skin) but they can be translated into insulation of the tissues in clo units, as in Fig. 21. The range

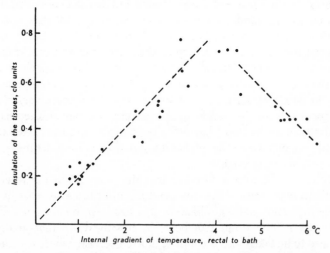

Fig. 21. Data of Burton and Bazett redrawn with clo units for insulation instead of in conductivity units. The fall of the insulation after the maximum corresponds to the increase in oxygen consumption (thermal tone and shivering of muscles).

of tissue insulation from full vasoconstriction to vasodilation is from 0·15 to 0·9 clo units. An unexpected and interesting finding was that there was a maximum insulation reached in cool baths. Still colder baths gave a considerable decrease from this maximum. The point at which the maximum occurred corresponded to the bath temperature below which increase of metabolic rate, by shivering, occurred. Evidently it is impossible that the heat production and muscular activity should increase without an increase of blood flow to the muscles. This means that shivering as a mechanism of defence against cold is inefficient, in that the cutting down of heat loss by full vasoconstriction is not possible when heat production is increased. Extra heat is produced but heat loss is greater than it was before shivering.*

Hardy and DuBois (12) also calculated the change in conductivity of the tissues, with the subjects in air instead of water. Their range of values for insulation of the tissues, three times instead of five times, is smaller than in the water bath experiments. However, Winslow *et al.* (13) find a change of four times between constriction and dilation.

There is a theoretical objection to considering the water bath experiments as typical of physiological reactions to temperature in air, since in well-stirred baths the skin temperature is forced to be the same (that of the water) over the whole of the immersed body. In normal environments of air the skin temperature differs greatly over the body. This difference means that the normal integration by the central nervous system of the afferent impulses from thermal receptors must be considerably interfered with in the bath experiments. There is evidence of this, in that a well stirred water bath that feels 'neutral', one of about 35°C (95°F), is one in which heat balance is far from achieved, and the body will steadily lose heat. To maintain balance the temperature of the bath must be at least a degree higher, and this initially gives an integrated sensation well on the warm side of neutral. However, data obtained on the range and limits for the thermal insulation of tissue with changes of peripheral blood flow remain applicable.

It turns out that in experiments on men in heavy protective clothing in very cold environments, there is a similar disturbance of integration of temperature sensations. In these circumstances

---

* A similar loss of thermal insulation of the tissues of the extremities occurs when the local protective 'hunting-reaction' occurs in the cold. This reaction is discussed in Chapter VIII.

the deep body temperature steadily falls to abnormally low values, 35·5°C (96°F), before there is any sensation of cold or metabolic response. The heavy clothing protects the skin temperature from falling to very low levels, while the very great external gradient of temperature from skin to environment makes the heat loss excessive. In the same group of experiments the values for the thermal insulation of the tissues were grouped about the mean value of 0·5 clo units (Fig. 22), instead of the maximum value of 0·8 clo

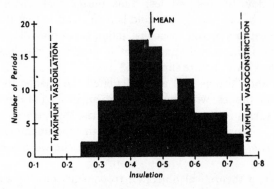

Fig. 22. Distribution of values for insulation of the tissues found in 80 one-hour periods on 40 subjects sitting heavily clothed in a cold chamber at temperatures from −7°C to −18°C (+20°F to 0°F) to illustrate the lack of vasoconstriction response, even though body temperature was falling rapidly.

units that would be found if vasoconstriction had been maximal (14). In colder environments (−29°C, −20°F) with a wind, the stimulation of the exposed face was sufficient to elicit the metabolic response and full vasoconstriction occurred very quickly, and these falls of deep body temperature did not occur.

A complication in using average skin temperatures over a large area must be considered. Love (15) showed that the heat loss of the hand might vary considerably even when the average temperature of the skin and the environmental temperature were unchanged, or their difference was unaltered. This apparently can result from a redistribution of blood flow, particularly to the fingers, and a shift of the venous return from superficial veins to the venae comites (16, 17). The latter are an apparatus for efficient heat exchange with the arteries which bring in hot blood. There must be similar changes in the insulation of the tissues resulting from such shifts in the route of venous return, and in the distribution of skin temperature. Due to the curvature effect, which is of considerable importance in the fingers, the effective insulation of the air is less for the fingers than elsewhere, and a shift in flow to the fingers will result in increased total heat loss, even though the average temperature of the skin may not increase.

Thus an important part of the physiological control of the total insulation of the tissues is the redistribution of blood flow to areas where the curvature effect alters the effective insulation, both of the tissues themselves and of the surrounding air. This consideration again emphasizes, as has already been discussed, how the gradient of temperature down the length of the limbs is more important in the control of heat loss of the animal than is the gradient from deep tissues to the skin lying over them.

The Relative Importance of the Various Insulations

Fig. 23 summarizes the last three chapters for the case of man, in that it represents the range of values usually obtainable in I_T, I_{cl}, and I_A. It is seen at once that in extremely cold environments

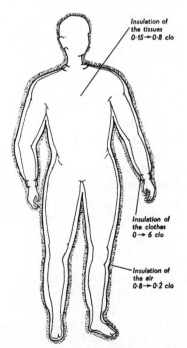

Insulation of the tissues
0·15 → 0·8 clo

Insulation of the clothes
0 → 6 clo

Insulation of the air
0·8 → 0·2 clo

FIG. 23. The various thermal insulations, and their range of values, that make up the total thermal insulation of man. (Burton, *Fed. Proc.*, **3**, 346, 1946.)

G

where the total insulation (the sum of I_T, I_{cl} and I_A) must be many clo units, the physiological control of I_T over 0·7 clo units does not appear to be of much consequence. However, there are areas of the body, like the face, which cannot conveniently be protected with clothing insulation (a satisfactory face-mask has not been designed so far), and the hands, where the total insulation is limited by the curvature factor, and here the physiological control of I_T remains of very great importance. Again, practical requirements demand that the clothing be reduced to the minimum that will give a long enough 'tolerance-time' in the cold. Where the insulation is only just sufficient to achieve heat balance, the state of the circulation and the value of I_T for the body that it determines will greatly alter the tolerance time, from an indefinitely long time if there is vasoconstriction, to a few hours if there is general vasodilatation. An example worked out by Burton and MacDougal (14) is given in the following table.

PREDICTED FALL OF MEAN BODY TEMPERATURE IN TOTAL CLOTHING INSULATION OF 4 CLO.

(Metabolism remaining at 50 cal/sq. m/hr)

Air temperature	0°C 32°F	−6·7°C 20°F	−20°C 0°F	−29°C −20°F	−40°C −40°F	
Fall of mean body temp. vasoconstriction	0	0·20	0·55	0·91	1·25	°C/hr
	0	0·36	1·0	1·64	2·26	°F/hr
Vasodilatation	0·12	0·37	0·77	1·15	1·55	°C/hr
	0·20	0·67	1·4	2·1	2·8	°F/hr

The vasoconstriction can therefore somewhat reduce but cannot prevent the fall of body temperature. Only an increase in metabolic heat can physiologically effect this. The table emphasizes the inadequacy of the insulation of clothing now available, when the wearer is inactive.

REFERENCES

1. BURTON. A. C. The Range and Variability of the Blood Flow of the Human Fingers. *Am. J. Physiol.*, **127, 4**37, 1939.
2. BORDIER, H. Thermal Conductivity of Tissues. *Arch. Paris, Physiol.*, **Serie 5 T 10**, 17, 1898.
3. LEFÈVRE, J. Studies on the Thermal Conductivity of Skin in vivo, and the Variations induced by Changes in the Surrounding Temperature. *Journ. de Physique.* June, 1901. See also LEFÈVRE, J. 'Chaleur animale et Bioenergetique,' p. 393–400. Masson *et al.* Paris, 1911.

4. HARDY, J. D. and SODERSTROM, G. F. Heat Loss from the Nude Body and Peripheral Blood Flow at Temperatures of 22°C to 35°C. *J. Nutrition*, **16**, 494, 1938.

5. HATFIELD, H. S. and PUGH, L. G. C. Thermal Conductivity of Human Fat and Muscle. *Nature*, **168**, 918, 1951.

6. BURTON, A. C. The Direct Measurement of Thermal Conductance of the Skin as an Index of Peripheral Blood Flow. *Am. J. Physiol.*, **129**, 326, 1940 (abstract only).

7. READER, S. R. The Effective Thermal Conductivity of Normal and Rheumatic Tissues in Response to Cooling. *Clin. Sci.* **11**, 1, 1952.

8. BAZETT, H. C. Discussed in his article in 'Physiology of Temperature Regulation.' p. 128–134. Newburgh, Saunders, Philadelphia, 1949.

9. BURTON, A. C. The Application of the Laws of Heat Flow to the Study of Energy Metabolism. *J. Nutrition*, **7**, 481, 1934.

10. BAZETT, H. C. and PETERSON, L. H. Theory of Reflex Controls to explain Regulation of Body Temperature at Rest and During Exercise. *J. Appl. Physiol.*, **4**, 245, 1951.

11. BURTON, A. C. and BAZETT, H. C. A Study of the Average Temperature of the Tissues, of the Exchange of Heat and Vasomotor Responses in Man of a Bath Calorimeter. *Am. J. Physiol.*, **117**, 36, 1936.

12. HARDY, J. D. and DuBois, E. F. Basal Metabolism, Radiation, Convection and Evaporation at Temperatures of 22° to 35°C. *J. Nutrition*, **15**, 477, 1938.

13. WINSLOW, C-E. A., HERRINGTON, L. P. and GAGGE, A. P. Physiological Reactions of the Human Body to Varying Environmental Temperatures. *Am. J. Physiol.*, **120**, 1, 1937.

14. BURTON, A. C. and MacDOUGALL, G. R. An Analysis of the Problem of Protection of the Aviator Against Cold, etc. Nat. Res. Council, Canada, Report No. C. 2035, 1941.

15. LOVE, L. H. Heat Loss and Blood Flow of the Feet Under Hot and Cold Conditions. *J. Appl. Physiol.*, **1**, 20, 1948.

16. BAZETT, H. C., LOVE, L., NEWTON, M., EISENBERG, L., DAY, R. and FORSTER II, R. Temperature Changes in Blood Flowing in Arteries and Veins in Man. *J. Appl. Physiol.*, **1**, 3, 1948.

17. BAZETT, H. C., MENDELSON, E. S., LOVE, L. and LIBET, B. Precooling of Blood in the Arteries, Effective Heat Capacity and Evaporative Cooling as Factors Modifying Cooling of the Extremities. *J. Appl. Physiol.*, **1**, 169, 1948.

CHAPTER 6

THE POSSIBILITIES OF MAINTAINING THERMAL STEADY STATE IN THE COLD, AND HOW ARCTIC ANIMALS DO SO

Now that the general equation of the thermal steady state of an animal has been formulated, it is easy to survey the methods by which the animal might maintain that state in the face of changes in the environmental conditions. The equation is

$$M = \frac{T_C - T_A}{I_T + I_{cl} + I_A} + E \quad \ldots\ldots\ldots\ldots\ldots 1$$

Suppose an animal is in the steady state at an environmental temperature T_A of 25°C (77°F), with its core temperature T_C at 37°C (98·6°F). The excess temperature $(T_C - T_A)$ is then 12°C (21·6°F). Let the environmental temperature fall by 12°C (21·6°F), i.e. to 13°C (55·4°F), so that the excess temperature is doubled 24°C (43·2°F). The heat loss by radiation and convection will tend to double. What factors in the equation may change to compensate for this? The possibilities are the following:

(*A*) *Compensation by Body Temperature.* The deep body temperature, or core temperature, T_C, of the animal might be allowed to fall correspondingly (i.e. to 25°C, 77°F), so that the excess temperature was once more only 12°C (21·6°F). The steady state would again be achieved without any change in the other parameters. This is very much what happens in the case of the poikilotherms, except that with the fall of body temperature the metabolism M is reduced, as are the rates of all chemical reactions. In the case of the homeotherms, however, this method of compensation is not possible physiologically. The efficient functioning of the whole organism depends upon the maintenance of a core temperature within narrow limits. The compensation must be by change in the other factors of the equation.

(*B*) *Compensation by Heat Production.* To match the increase in the excess temperature, and the doubling of the second term on the right-hand side of the equation, there might be a corresponding increase in heat production. It turns out that this is indeed the final method of compensation in the homeotherms, but is utilized only when other methods of compensation have reached the limit of their effectiveness.

(C) *Compensation by Insulation.* The increase in the numerator of the fraction, i.e. in the excess temperature, might be met by a corresponding increase in the denominator, i.e. in the total thermal insulation. Obviously under this heading would be increases in any or all of the components of the total thermal insulation, in the insulation of the tissues I_T, of the fur or clothing I_{cl}, or of the air I_A. Increase of thermal insulation is the method of regulation most employed by the homeotherms, though its scope of effectiveness is quite severely limited in some cases.

(D) *Compensation by Evaporation.* There might be a change in the evaporative loss E to compensate. While this is the main method of compensation against a rise of environmental temperature, by increase of evaporative loss (sweating, panting), its usefulness in cold conditions is obviously very limited. When sweating or panting is not present, the evaporative loss of heat amounts to only about 25 per cent of the total heat loss, so even if this were reduced to zero, compensation for increases in excess temperature of more than 25 per cent of the original value would be impossible. Not even this, however, is possible, since the evaporative loss cannot fall below the minimal value of the 'insensible loss'. For example, in men at rest, the insensible loss falls only by a few per cent in environments from 20°C to 10°C (68°F to 50°F).

Of course, a combination of these compensating factors may operate to bring about a new steady state. In the equation is indicated what combinations will be necessary. A survey of what methods of compensation are actually used by animals in cold environments is most useful, since it permits us to explore the possibilities of further exploitation of methods of compensation, which have not been fully utilized by animals. Thus the directions of applied physiological research which will be profitable, or profitless, are seen. Such a survey is made possible by the extensive work of Irving and Scholander *et al.* (1) (*a*) and (*b*) on Arctic, temperate and tropical animals.

Possibility (A), the compensation by allowing a change in the body temperature, has been dismissed as inapplicable to the homeotherms. It apparently has not been a factor in the adaptation of species to the climate of their habitat. Comparing similar species of birds and mammals, the one living in the tropics, the other in the Arctic, there is no consistent difference in the temperatures at which they regulate the deep tissues. In man there is certainly no evidence that the Eskimo has a 'core temperature' any lower than the native of the tropics.

Heat Production and 'Chemical Regulation'

Possibility (*B*) of compensation by increasing the heat production is universally employed in thermal regulation of animals when the environmental cold is great enough. Indeed it is ultimately the only remaining defence of the animal against loss of body temperature. In lower animals, where the regulation is not nearly so well

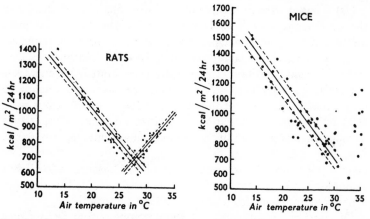

FIG. 24. Metabolic response to temperature in small laboratory animals. (From Herrington. Am. J. Physiol. **129**, 123, 1940.)

developed and the 'core temperature' is quite variable, as in the bat (2) and the sloths (3), the compensation by increase of metabolism would appear to be almost the only method of adjustment. This is well illustrated for the albino rat by the work of Herrington (4). As Fig. 24 shows there is only a very narrow range of environmental temperature over which the metabolic rate is constant. Below this range it increases as a compensation to the increased heat loss; above this narrow range the metabolism also increases, presumably in obedience to the thermo-chemical law. Mice and guinea pigs have only a slightly greater range of 'environmental neutrality'.

Recourse to increased metabolic rate, and heat production against cold, obviously involves some sacrifice of the freedom of the animal economy. Normally the metabolic rate is varied in accordance with the purposeful muscular activity of the animal, and is subject to limitations as to the availability of food. If metabolism has to be maintained at a high level determined by the demands of temperature regulation, this can be regarded as a restraint upon the animal. In animals such as the dog, cat and man, which have a more highly developed thermal regulation, other methods of compensation, by insulation and by evaporation are effective over a wider range of environmental temperature. Below this range, they also have to resort to increase in heat production.

The climatic adaptation of species shows itself most dramatically in this matter of the dependence upon compensation by heat production. The Arctic animals do not have to resort to increased metabolism until the environmental temperature is very low, while

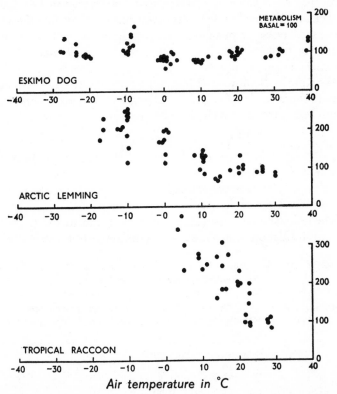

FIG. 25. Metabolic response to cold of two Arctic and a tropical animal. The low 'critical temperature' below which heat production increases for the Arctic animals is to be noted. (From Scholander *et al. Biol. Bull.*, **99**, 259, 1950.)

their counterparts living in the temperate or tropical regions must do so in response to relatively small falls in environmental temperature. A striking example from the work of Irving and Scholander is the Arctic white fox, which does not increase its metabolic rate unless the temperature is below −40°C or F. Similarly the Eskimo dog has a 'critical temperature', below which its heat production begins to increase, of from 0°C to −10°C (32°F to 14°F), the Arctic lemming of about −12°C (+10°F) and polar bear cubs about 0°C (32°F). In contrast the tropical raccoon has a critical temperature of +25°C (77°F). Fig. 25 illustrates some of these data.

Amount of Increase of Heat Production Required

The key to this remarkable adaptation in Arctic mammals and birds has been shown by Irving and Scholander to lie in their thermal insulation. It follows from our fundamental equations that the effectiveness of metabolic compensation, as well as the temperatures below which it will be needed, is determined by the thermal insulation.

Working in clo units and Met. units, we find from the equation that if there is a 10°C (18°F) change in environmental temperature T_A, the increase in M to compensate is given by

$$\triangle M = \frac{5 \cdot 5 \times 10}{I_T + I_{cl} + I_A} = \frac{55 \text{ kcal/sq. m/hr}}{I_T + I_{cl} + I_A} = \frac{1 \cdot 1 \text{ Met.}}{I_T + I_{cl} + I_A}$$

The increase in metabolic heat required to compensate each 10°C (18°F) fall in environmental temperature is inversely proportional to the total insulation of the animal.

The following table is made up to show this feature of the use of thermal insulation, which has not had the notice it deserves.

Increase in Metabolic Rate Required to Compensate for Fall of Environmental Temperature

Total insulation of animal	$\triangle M$ for 10°C drop in environmental temp.		Drop in environmental temp. requiring 1 Met. extra metabolism	
clo units	kcal/sq.m/hr	Met. units	°C	°F
0·5	110	2·2	4·5	8
1·0	55	1·1	9	16
2·0	28	0·56	18	32
5·0	11	0·22	45	81
10·0	5·5	0·11	100	180

Thus an unclothed man in still air, with a total insulation of 1·6 clo ($I_T = 0·8$, $I_{cl} = 0$, $I_A = 0·8$) will have to double his resting metabolic rate for each 8°C or 15°F. In a high wind ($I_A = 0·2$) he will have to double his metabolism for each 5°C or 9°F fall of temperature. In contrast in a heavily clad man or furred animal, where the total insulation is 6 clo units, doubling the metabolic rate would compensate for a drop of 26°C or 46°F. Indeed it can be estimated that for the Arctic fox, which does not have to call upon increased heat production until the temperature is below −40°C or F, doubling its metabolism would enable it to maintain a thermal steady state at −120°C (−184°F). It takes only a 37 per cent increase to sustain it at the lowest temperatures recorded on earth (about −70°C or −85°F). The total insulation of the Arctic fox must be about 10 clo units.

Limit to Metabolic Compensation

It has been shown for man (5) that in the most violent shivering the metabolic rate does not exceed about 3 Mets.* and this would appear to be also the approximate maximal value in animals also, from a survey of the curves in the literature (1 and 4).

Let us use this maximal rate of 3 Mets. to calculate the lowest environmental temperature that could be sustained indefinitely by unclothed man. Putting 3 × 50 kcal/sq. m/hr for M and I_T = 0·8 clo (the value of tissue insulation for full vasoconstriction), $I_A = 0·5$ clo units (the average for outdoor conditions), we find from Equation 1 that the maximal gradient ($T_C - T_A$) will be about 35°C (63°F). The steady state could be maintained, even with continuous maximal shivering, only down to about 2°C (35·6°F) or very near freezing temperatures. This limiting temperature for unclothed man agrees very well with the environmental temperature known to apply to the Tierra del Fuegans, who endured it naked. Possibly extra layers of fat, in their case, increased the value of I_T.

All this emphasizes that compensation by increased heat production is not only restrictive to the freedom of life of the animal, but is dependent upon thermal insulation for its effectiveness, and

* Swift in his summary states that the metabolic rate can increase by 400 per cent, but this is a calculated value based on an estimate of the percentage of the time there was actual shivering. No steady rates exceeded 3 Mets. There is no doubt that in man, as in other animals, rates of heat production by shivering can exceed 5 Mets, but only for brief periods. (See Chapter 9.)

without great thermal insulation it is quite limited in its range of possible usefulness. An additional point has already been made in the last chapter, that metabolic response suffers a loss of efficiency in that increased heat production also involves a decrease in tissue insulation and thus an increase in heat loss.

The remarkable results of Scholander *et al.* for mammals are summarized in Fig. 26. A similar diagram describes the metabolic compensation to cold of Arctic, temperate and tropical birds.

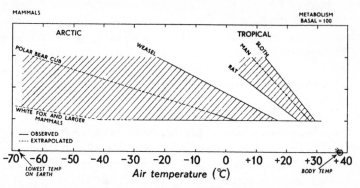

Fig. 26. Summary diagram from the work of Scholander *et al.* The rise of heat production below the 'critical temperature' is in all cases described by a straight line which if extrapolated would intersect the axis at the body temperature (37–38°C). (Scholander *et al.*, *Biol. Bull.*, **99**, 259, 1950.)

Compensation by Insulation

An increase in the total thermal insulation could be brought about by an increase in any of the components, i.e. the tissue insulation, the insulation of fur or clothing, and that of the air.

The limits of compensation by increasing the tissue insulation are soon reached, for the full range of physiological alteration appears to be, in man, from 0·15 clo for full vasodilatation to about 0·8 to 0·9 clo for full vasoconstriction. It is possible that increases of thickness and changes in quality of subcutaneous fat could further raise the maximal value, but it is difficult to see how this could result in appreciable increases (see Chapter 10 for further discussion).

In the case of the marine mammals, like the seal and the whale, where according to the *Encyclopaedia Britannica* the thickness of blubber can be up to 14 in. in the sperm whale and up to 20 in. in the Greenland whale, adaptation to cold by increase of adipose tissue can be an important factor. The thickness of blubber is so great that it may provide all of the thermal insulation necessary to maintain the steady state in cold water, down to freezing temperatures, without great increases of metabolism. In the case of man, however, where the surgeons encounter thicknesses of adipose tissue up to an inch on the trunk, and much less on the limbs, the importance of tissue fat in providing compensation against cold environments may be questioned. The insulating power of such fat layers could hardly be more than 1 clo/in. It is an undoubted fact that some men acclimatized to immersion of the hands in cold water can endure this for long periods without the tissue damage that would occur in those not acclimatized. We must look for the explanation elsewhere than in increase of insulation of the tissues by subcutaneous fat.

The relatively small range of physiological adjustment of the tissue insulation, even with adipose tissue, has another connotation. There is evidence that the acclimatization of the extremities mentioned above is connected with the curious phenomenon of local protective vasodilation in the cold (see Chapter 8). This may keep the temperature of the local tissues from falling to the point where tissue damage will occur, but it represents a reaction which is contrary to the thermal economy of the whole organism, in that the total tissue insulation must be thereby lowered. At external temperatures near or below freezing, where this phenomenon is seen, the loss by vasodilation of half a clo unit in the total thermal insulation of the animal is not great in comparison to the many clo units of insulation required by the animal in these circumstances. The interests of the whole organism are to some extent sacrificed to save the individual members, but the sacrifice is not so great as seriously to affect the thermal outcome for the whole animal.

As already discussed, the great adaptation in Arctic mammals appears to be by means of the thermal insulation of the fur. Fig. 16, used in an earlier chapter in another connection, shows this clearly. By analogy, we see that to enable man to withstand low environmental temperatures we must turn to the thermal insulation of clothing to provide the bulk of the compensation. As with fur, clothing will not only reduce the 'critical temperature' below which there will have to be a metabolic increase, but also will make the magnitude of the increase required much less than without the insulation of the clothing.

The finding that the compensation in Arctic mammals is almost entirely by increase of thermal insulation, while it greatly illuminates the biological field, raises at once a new problem, which certainly has not yet been solved. If the Arctic fox has a total insulation, mostly in its fur, of 10 clo units, which keeps the heat loss normal at $-40°$C or F, how does the animal manage to lose enough heat at higher temperatures such as $+15°$C$(59°$F)? With the same total insulation the heat loss would be reduced proportionally to the total gradient, now $22°$C $(39·6°$F) instead of $77°$C $(138·6°$F), so the heat loss would be, if no compensating factor came into play, only $\frac{2}{7}$ Met. One possibility might be in an increased evaporative loss at higher temperatures. An increase of $\frac{5}{7}$ Met. by evaporative loss is well within the limits of evaporative regulation. However, it would be supposed that the Arctic fox, like the dog, does not possess sweat glands distributed over the skin surface as in man, and such increased evaporative loss would have to be by panting. Apparently measurements of evaporative loss have not been made, but observation of the behaviour of these animals gives no evidence of such panting at temperatures above $-18°$C$(0°$F). One is forced to conclude that the animal has some way of effectively reducing the very great over-all insulation provided by the fur. Pilomotor reflexes which might increase or decrease the thickness of the fur may play a part, but experiments on the fur of a lemming, fluffed up and smoothed down,* showed a change of insulation of only twice, while a decrease of 10 or 11 times would be required to adjust to the higher temperatures (1). The only possibility of explanation seems to be that the effective decrease in insulation is achieved by a shift of the heat loss from well insulated parts of the body, like the trunk, to the poorly insulated parts like the legs. This could be accomplished partly by changes of posture and partly by redistribution of peripheral blood flow, as to the legs. It is likely that remarkable vasomotor shifts of this kind could be demonstrated in Arctic animals, if research were directed to this feature.

The example of the Arctic fox should emphasize the need, in protecting the human by thermal insulation, for the provision of adaptability in clothing insulation. Even where the environmental temperature remains low, the great changes in heat production due to activity demand that Arctic clothing be capable of reduction in insulation rapidly and easily. It has already been stated, in Chapter 4, that there is some evidence that conventional Arctic clothing

---

* The behaviour of the cat in hot weather in smoothing down the fur, with its paws, wet with saliva, is an interesting example of regulation.

suffers an automatic and considerable decrease in insulation during exercise, in which such a decrease is desirable. This is probably due to the increased penetration of air and increase in air movement within the clothing. Research in clothing design should be directed towards increasing this tendency, rather than decreasing it, and new methods of providing reduction at will of thermal insulation of clothing must be sought.

Just as increased heat production entails a change in insulation, (because of the decrease in tissue insulation with increased blood flow to the muscles), so increased thermal insulation entails some change in the metabolic activity of the animal. This is because the extra weight of the fur or clothing must be carried by the animal, and its bulk makes movement of the limbs somewhat more difficult so that more energy is required for the movement. Measurements of the increased cost of exercise with heavy clothing have been made (6). The extra cost is astonishingly low, amounting to only 5 per cent even with heavy clothing. While the change in tissue insulation involved in shivering was disadvantageous to the thermal economy, this effect on heat production of heavy clothing is, of course, in the direction of improved thermal economy, though it decreases slightly the unique advantages of compensation by insulation.

Compensation by Change of I_A

It must not be overlooked that the animal can effect an increase in the insulation of the air, even though this is primarily determined by the wind velocity. This can, of course, be done by the use of shelter from the wind, as with Eskimo dogs whenever they lie in the snow. It can also be accomplished by the adoption of a posture which reduces the surface area exposed to the wind to a minimum, as does the curled-up position of animals at rest in the cold. There is a further way in which the effective average insulation of the air for the whole body can be increased. The curvature effect, which results in the insulation of limbs of small radius of curvature being much reduced, has already been explained. If there be a physiological shift of blood flow away from such regions of small radius of curvature, the limbs and digits, to the parts of the body which are large, as the trunk, the average insulation of the air will be increased. Such a shift from flat to curved areas seems to take place in the hand, where the fingers, which have a higher skin temperature than the rest of the hand in hot surroundings, become much cooler than the rest in the cold. The effective average insulation of the air for the whole hand is thus reduced (7). However, the maximum increases in the thermal insulation of the air that can be produced by the physiological or behavioural reactions of the animal, other than by finding enclosed shelter, are not very great, being certainly less than 0·6 clo units.

Nervous Mechanism of Physiological Defence Against Cold

This has been so well treated in so many textbooks and articles (8), that these should be consulted. The mechanism of temperature regulation provides an example of co-ordinated reflex action, involving almost all of the functions of the organism, which is unparalleled. Fig. 27 may be useful in summarizing the range of physiology that is involved.

As with all reflexes there are the receptor mechanisms, the afferent pathways, the co-ordinating centres, the efferent arcs, and the effector mechanisms to consider.

The receptor mechanisms are considered by some to be dual. The chief of these are the temperature sense endings distributed over the surface of the body. However, even if the skin be kept warm, the reactions of defence occur when the 'core temperature' is sufficiently lowered (9). This might be interpreted as indicating a temperature receptor of some kind in the brain, affected by the temperature of the brain blood stream. This is probably too crude a concept. The suggestion of Bazett (8) should be much more acceptable to physiologists. This is that the sensitivity or the relaying and co-ordinating function of the hypothalamic centre, or rather 'centres' since there is definite evidence of the anatomical and physiological separation of the heat loss and heat conservation centres (10), is affected by the temperature of these centres. If the optimal temperature, at which afferent impulses from cold receptors on the skin evoke a maximal response from the heat conservation centre, were low, say 36°C (96·8°F) then a fall of brain blood temperature would result in increased reflex activity against cold, even if the afferent stimulation were unchanged. Since the two centres are most probably mutually inhibitory, a fall in brain temperature would thus also depress the activity of the heat loss centre, or this centre might have its optimal temperature for response higher than normal body temperature. Thus there is no need to postulate a specialized 'thermal receptor' in the brain, for which we have no histological evidence whatever.

The effector mechanisms include practically all the known motor functions of the tissues and all of the organs, save those of reproduction. Further detail in the diagram, or in discussion here, would only reduce any usefulness it may have.

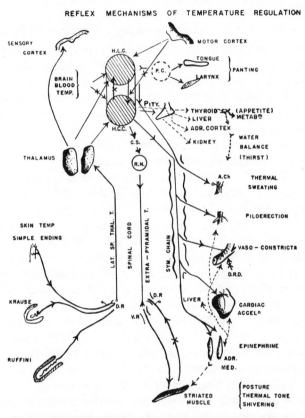

FIG. 27. Reflex mechanisms of temperature regulation. Schematic diagram to summarize the reflex control of body temperature. The arrows are not to be taken as indicating single neurons but merely the nervous pathways.

From the point of view of cybernetics, it is of great interest that the body regulates its heat loss, which is the product of the thermal gradient and the effective thermal conductivity of the tissues, by controlling this conductivity according to the gradient. Indeed, there is much evidence that the temperature receptors in the skin respond to the thermal gradient in their locality rather than to their temperature (8). Such a regulatory system has not been commonly utilized by engineers, though it would appear to be the soundest method theoretically.

Cybernetic Considerations, Man as a Tropical Animal

With the increase in general interest and knowledge of controlling or 'servo' mechanisms, 'cybernetics', it becomes useful to think of physiological regulating systems in the engineering terms employed in that field of study. Some considerations of the mechanisms of temperature regulation of animals in these terms have already been published (11).

Any co-ordinating mechanism which integrates a large number of diverse 'afferent' signals must of necessity operate best when there is a certain 'normal' relation between the environmental factors which elicit these afferent signals. Thus the integration by the brain of afferent temperature stimuli from the receptors situated all over the body, must be based upon a normal distribution of such stimuli from the different areas. The skin temperature in comfortable indoor conditions (21°C (70°F) with light clothing) varies greatly over the body, and so does the density of population of the thermal receptors. It is not surprising therefore to find that when this normal distribution of afferent stimuli is disturbed, as in a well-stirred water bath where the skin temperature everywhere on the body will be brought to equality, the integrating control mechanism will be less efficient than in the normal environment of air. Thus (12) a bath of temperature about 35°C (95°F) will be 'thermally neutral' in sensation to an immersed subject, and there will be no compensating thermoregulatory reaction by the body. As a consequence the body will steadily lose temperature in such a bath, until the core temperature has dropped sufficiently to elicit a response. To maintain the thermal steady state the bath must be a degree or so warmer than 35°C (95°F), and this will feel distinctly on the warm side of neutral. Another example of this interference with the optimal nervous integration has also been mentioned in an earlier chapter. Wearing of heavy clothing in the cold maintains a fairly high average skin temperature, while the gradient of temperature from skin to environment is so great that, in spite of insulation of the clothing, the heat loss is excessive. The physiological regulatory

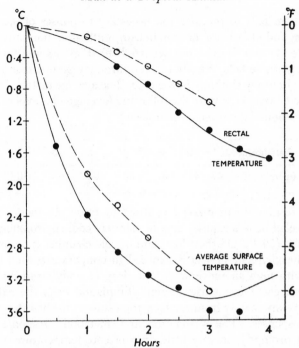

FIG. 28. Fall of body temperature in flying clothing of 2 to 2·5 clo at 0° to 10°F. Broken lines −3·0 clo. Mean values for 40 subjects.

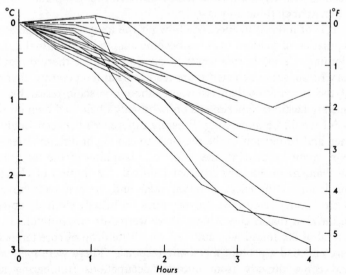

FIG. 29. Fall of rectal temperature. Data for individual subjects.

H

mechanism fails to produce the necessary responses of vasoconstriction and of increase of metabolism, until an abnormally great
fall of body temperature has occurred (9). Figs. 28 and 29 show
how large these falls of body temperature may be in a man seated
quietly, in heavy clothing, in the cold. The average metabolic rates
in these experiments were the following (average of 8 experiments
each of 4 hours' duration, on 4 subjects).

Time in half-hours	l	2	3	4	5	6	7	8
Cal/sq. m/hr	45	44	46	56	64	63	70	81

These results are in marked contrast to those obtained when a
lightly-clad man is exposed to a moderately cool environment, say
of 15·5°C (60°F). On the other hand in cold chamber experiments,
if the heavily-clad subject is exposed to temperatures well below
zero, with high air movements, the drop of body temperature is
seldom seen. The intense afferent stimulation from the exposed
areas of skin, mainly from the face, then suffices to elicit the necessary responses. The vasomotor and metabolic responses are
elicited promptly, so very little drop of core temperature occurs.
The integrated regulatory mechanism is evidently adjusted for
dealing with environmental temperature in the 'temperate' range,
when light clothing only is worn.

It is of interest to speculate how far the integrative temperature
regulation of modern man has become adapted to the usual indoor
clothing. Certainly it is common opinion that the thermo-regulatory responses are elicited more readily from temperature changes
of the extremities than elsewhere, though specific researches on
this are lacking. Experiments on habitually unclothed men on this
point would be interesting. Possibly differences between English-
men and Americans in the response to cooling of different areas of
skin might be evident, the result of adaptation of the integrated
mechanism to the very different habitual distribution of clothing
insulation. Differences between men and women have already
been demonstrated (13). In the series on heavily clothed subjects
in the cold that is cited above, there were only two individuals, of
a total of 60, found who suffered very little drop of core temperature and had a prompt metabolic response. These were men who
had come directly from outdoor occupations (lumbering and
transport driving) in the Canadian winter. More data on this are

Man as a Tropical Animal

105

needed, to show whether or not there is an acclimatization to cold in man involving an adaptation of the integrative nervous mechanisms of the kind discussed.

The work of Scholander and his colleagues (1) suggests to them that 'man is a tropical animal', and his physiological mechanism of homeostasis of body temperature is essentially adapted to tropical rather than cold climates. He may be enabled to live in the cold by wearing heavy clothing, but his physiological mechanisms will be at a disadvantage. The physiology of a heavily clothed man in the cold is probably as much disturbed as would be that of an Arctic mammal in warm environments if we clipped off his heavy coat of fur.

REFERENCES

1 (a). SCHOLANDER, P. F., HOCK, R., WALTERS, V., JOHNSON, F. and IRVING, LAURENCE. Heat Regulation in some Arctic and Tropical Mammals and Birds. *Biol. Bull.*, **99**, 225, 1950.
   (b). SCHOLANDER, P. F., HOCK, R., WALTERS, V. and IRVING, LAURENCE. Adaptation to Cold in Arctic and Tropical Mammals and Birds in Relation to Body Temperature, Insulation and Basal Metabolic Rate. *Biol. Bull.*, **99**, 259, 1950.
2. HOCK, R. Temperature and Bat Metabolism. Unpublished doctoral thesis. Cornell Univ. 1949.
3. WISLOCKI, G. B. and ENDERS, R. K. Body Temperature of Sloths, Anteaters and Armadillos. *J. Mammal.*, **16**, 328, 1935.
4. HERRINGTON, L. P. The Heat Regulation of Small Laboratory Animals at Various Environmental Temperatures. *Am. J. Physiol.* **129**, 123, 1940.
5. SWIFT, R. W. The Effects of Low Environmental Temperature upon Metabolism. 2. The Influence of Shivering, Subcutaneous Fat, and Skin Temperature on Heat Production. *J. Nutrition*, **5**, 227, 1932.
   See also DuBois, E. F. 'Basal Metabolism in Health and Disease'. Lea and Febiger, Philadelphia, 1927.
6. GRAY, E. LeB, CONSOLAZIO, F. C. and KARK, R. M. Nutritional Requirements for Men at Work in Cold, Temperate and Hot Environments. *J. Appl. Physiol.*, **4**, 270, 1951.
7. DAY, R. 'Regional Heat Loss'. Chapter 7 in 'Physiology of Heat Regulation and the Science of Clothing'. Newburgh, Saunders, Philadelphia, 1949.
8. BAZETT, H. C. 'The Regulation of Body Temperatures'. Chapter 4 in 'Physiology of Heat Regulation and the Science of Clothing'. Newburgh, Saunders, Philadelphia, 1949.
9. BURTON, A. C. and MacDOUGALL, G. R. An Analysis of the Problem of Protection of the Aviator Against Cold and the Testing of the Insulating Power of Clothing. Nat. Res. Council, Canada, Report No. C.2035, 1941.
10. BLAIR, J. R. and KELLER, A. D. Complete and Permanent Elimination of the Hypothalamic Thermogenic Mechanism Without Affecting the Adequacy of the Heat Loss Mechanism. *J. Neuropath. and exper. Neurol.*, **5**, 241, 1946.

11. BURTON, A. C. 'The Operating Characteristics of the Human Thermo-regulatory Mechanism. Temperature, its Measurement and Control in Science and Industry' (Symposium of the Am. Inst. of Physics), p. 521, Reinhold, New York, 1941.
12. BURTON, A. C. and BAZETT, H. C. A Study of the Average Temperature of the Tissues, of the Exchanges of Heat and Vasomotor Responses in Man by Means of a Bath Calorimeter. *Am. J. Physiol.*, **117**, 36, 1936.
13. HARDY, J. D. and DuBois, E. F. Differences in Men and Women in Their Response to Heat and Cold. *Nat. Acad. of Sci.*, **26**, 389, 1940.

CHAPTER 7

THE ESTIMATION OF THE THERMAL DEMAND OF THE ENVIRONMENT

Popular Knowledge and Precise Equations

The man in the street is thoroughly familiar with the prediction of his thermal comfort based upon the level of the environmental temperature alone. The expected temperature is daily published in the newspapers, usually in a place of honour, and a greater aggregate of time 'on the air' probably is given to reports of what the temperature was, is, and is likely to be, than to any other single topic. Most of us, without analytical thought, have learned to estimate, from the temperature of the environment alone, roughly how great will be the heat loss from the human body, or more usually we estimate how much thermal insulation in the form of clothing we will need to reduce that heat loss to within normal limits. The equations already given by which the heat loss can be calculated from the external environmental temperature, and from the thermal insulation of the clothing and that of the air, are thus merely more precise expression of a very widespread intuitive public knowledge.

The 'man in the street' is apt to translate this knowledge of the relation of temperature to thermal demand in terms of the amount of clothing he will need to wear to maintain comfort, whether he should wear his overcoat or not, whether the light or the heavy overcoat, and so on. On this basis, it might be decided that the precise estimate of thermal demand of the environment would be best expressed as the clo units of insulation required to maintain comfort in that environment. The equations for doing this are already provided in our simple scheme, in which first the total insulation, of clothing plus air, is calculated. The graphs for total insulation required are now very well known (Fig. 30).

The solicitous wife of the man in the street often goes a step further in unconscious analysis of the factors of thermal demand, in that in addition to taking into account the environmental temperature registered by the thermometer, outside the kitchen window in most Canadian homes, she recognizes the rôle of air movement. In urging that the heavy, rather than the light overcoat be chosen, she will point out that, though the thermometer may indicate 4°C (40°F), there is a 'cold wind'. Our scheme takes this factor into quantitative account in that from the total insulation required the insulation of the air is subtracted, and this value is dependent upon the velocity of air movement, as in those tables and graphs already given.

Fig. 30 shows that there is not one straight line, but a whole

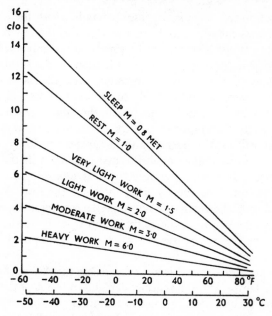

FIG. 30. Total insulation, of clothing plus air, needed for different metabolic rates. The formula for the lines is:

$$I = 0 \cdot 082(91 \cdot 4 - T°F)/M$$

Total insulation $I_{cl} + I_A$ for different environment as temperatures and metabolic rates. For the insulation of clothing required, I_{cl}. The insulation of the air I_A, dependent on the wind, must be subtracted from the total insulation.

family of lines, one for each 'metabolic rate'. Popular analysis has omitted a most important factor. The thermal insulation required for comfort depends to a very great extent upon the activity of the man exposed to that environment. Rather successful attempts to make this obvious to the military 'layman' who might sit upon committees concerned with provisioning of troops in different climates were by means of pictograms (Fig. 31, from the Medical Art Department of the University of Toronto). Again it must be

How many clo units needed?

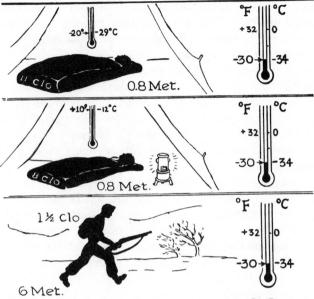

What 1 clo unit is good for.

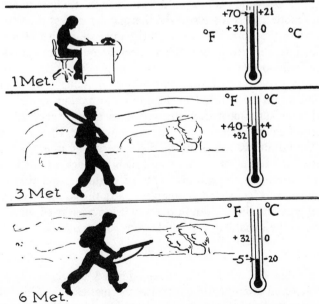

FIG. 31. Pictograms to illustrate the importance of the metabolic rate in determining the clothing required in different environments.

admitted that the really thermally well-educated and solicitous wife is not unaware of this all-important metabolic factor. She will advocate a rug in addition to the overcoat if the subject is to be sitting watching a football game.

Though the simple quantitative scheme, taking into account the environmental temperature, the air movement, and the metabolic rate, is still sound and scientific, though subject to the inevitable defect of being an 'average' expression for the whole body where the requirement of different parts of the body are very different, it is no longer very satisfying. Already we must make a double modification of our estimate of thermal demand which was based originally upon the environmental temperature alone. When it is further realized that whether or not the sun is out, i.e. the intensity of solar radiation or 'insolation', also makes a good deal of difference, and so, in a mysterious way, does the 'dampness' of the air, i.e. the humidity, the inclination becomes strong to sacrifice some precision, if necessary, to achieve a simpler mode of estimation.

Unitary Estimates

There are several classical and some newer attempts to achieve unitary modes of estimation of the thermal demand of the environment upon the human body. It is the purpose of this chapter to inquire into the possibility that these unitary schemes might be scientifically sound.

It is an attractive idea that simplicity of thinking about thermal demand might be reached by having a 'correction factor', to be added to or subtracted from the environmental temperature. This correction would depend upon the air movement, the solar radiation, the humidity, and even upon the metabolic rate. Having so corrected the temperature, we could then fall back on our familiar way of estimating the requirements of clothing from this single parameter, the 'corrected temperature'. It turns out that this is indeed possible, though not in the way it has been tried, in respect to some of the variables, impossible in respect to others.

Wind-chill

The table of 'wind-chill values' devised by Siple (1) is an example of an attempt to include the thermal effects of air movement with those of temperature in a single index. Here the result was expressed not, however, in the form of a correction to the temperature for the wind, but in a new scale of rate of heat loss, said to be kcal/sq. m/hr, running from 50 (hot) to 1,000 (very cold), to 2,500 (intolerably cold and the maximum 'cooling rate' recorded under natural outdoor conditions). This index is read from charts, or 'nomograms', of the two variables, namely the wind speed and the external temperature.

The index of 'wind-chill' has enjoyed a considerable, and deserved, popularity, for it has been proved in the field that it does indeed provide an index corresponding quite well with experience in the cold, i.e. of the discomfort and tolerance of man in the cold. Yet it can easily be shówn that the scientific basis for 'wind-chill' is lacking, so that its empirical success becomes a matter for investigation.

Why has 'wind-chill' no scientific basis? Because the heat loss depends upon the 'total' insulation, made up (assuming a standard skin temperature for comfort, so we do not have to consider the insulation of the tissues) of the sum of the clothing insulation and the insulation of the air. The degree of air movement affects primarily only one item of the total, that is the insulation of the air. It is true that a secondary effect, if the clothing is not adequately protected by wind-proof covers, will be an effect on the insulation of the clothing, but this does not affect the argument that it *is theoretically impossible to express the effect of wind on heat loss without references to the amount of clothing that is being worn.* When it is realized that the maximum effect of a wind on the insulation of the air is about 0·8 clo (I_A is 1·0 clo for very still air and 0·15 for a very high wind), and in the total insulation up to 5 clo units of insulation of clothing may be added, it is seen that the amount of clothing worn is not to be considered as a small modifying factor to wind chill values, but theoretically should be the dominant factor in determining the heat loss. There ought to be, instead of one chart for calculating 'wind-chill', a set of widely differing charts, one for each value of clothing insulation worn.

This objection is not merely theoretical, for it is easy to show experimentally that the same wind speed may increase the heat loss of a lightly clad man very greatly, but increases only slightly the heat loss of a heavily clothed man. In a series of experiments to check this, done in the R.C.A.F. cold chambers in Toronto in 1942, subjects were exposed to a 12 m.p.h. wind, from a fan, in light clothing, about 1 clo unit, and their heat losses were measured in comfortable indoor conditions. These increased about

50 per cent in the wind over that in still air. The same subjects were then heavily clothed (4 clo units), and exposed in the cold chamber so that they were at a comfortable temperature, again with and without the same wind. In this case their heat loss was increased only about 12 per cent. These rates of increase are utterly out of line with those predictions from 'wind-chill' charts, but in agreement with the predictions of our scheme of I_{cl} plus I_A.

How then can we explain the proven usefulness of the 'wind-chill' index in practice? The answer is not far to seek. Experience in the cold will verify that the tolerance of a man is finally determined by the parts of his body which usually are unprotected, such as the face and hands. No really satisfactory face-mask for the functioning soldier has been available and gloves have often to be removed for fine work. For these parts the total insulation is that of the air only, and the values of 'wind-chill' are based upon the observations made by Siple, of the rate of cooling of 'naked', uninsulated, cans of water. The 'wind-chill' should then correspond with the cooling of the naked body, or of those parts of it which are unprotected even in the heavily clothed man.

The 'wind-chill' index should certainly be retained as a proven practical guide to tolerance in the cold, but it may be pointed out that the invention of a really satisfactory face-mask (and this should be of high priority), or other ways of protecting parts of the body at present exposed, would make it obsolete. The tables of 'wind-chill' would then greatly exaggerate the importance of wind in increasing the total heat loss, as they do now, and the shortening of the tolerance time, for which they are now empirically adequate.

The Thermal-Wind-Decrement

Fortunately in the case of this additional variable, the air movement, there is theoretically a simple way to express its effect on the thermal demand as a correction to the environmental temperature. It is relatively simple, but the amount of the correction must depend upon the metabolic rate of the individual, so separate tables of the correction for men engaged in different activities are needed. Since it is impossible to include the metabolic rate in a unitary scheme, we must continue to have separate standards for men of different degrees of activity, but this is no real disadvantage.

If we are able to express the effect of wind as a correction, or decrement, to the temperature of the air, we can then use this corrected temperature to predict the thermal demand from the knowledge of how this would alter with the temperature in still air. Such a corrected temperature would then be called the 'equivalent still-air temperature'.

We start devising such a correction for wind to the temperature by expressing the effect of wind on the thermal insulation of the air in a similar manner, i.e. as a decrement of insulation, 'W', to be subtracted from the insulation of still air. This decrement W is very easily calculated from the table of I_A for different wind velocities already given (Chapter 3) so that:

$$I_A = I_{S.A.} - W$$

Insulation of the air. clo units	1·0	0·8	0·7	0·6	0·5	0·4	0·3	0·2	0·1
Ft./min.	20	35	50	75	120	210	425	1,050	4,500
Miles/hr	0·23	0·40	0·57	0·85	1·37	2·39	4·85	11·9	51·0
Insulation decrement W clo	0	0·2	0·3	0·4	0·5	0·6	0·7	0·8	0·9

For the standard value for still air $I_{S.A.}$ we might choose any convenient value of air movement commonly encountered in still air such as in a room without draughts. Random air movement is seldom less than 20 ft./min. even in such a room, and this allows us to choose 1·0 clo units as our standard value for $I_{S.A.}$, from which the decrement W is calculated.

For a thermal steady state, using clo units for insulation and Met. units for heat production or loss,

$$M - E = 0·11 \times \frac{T_S - T_O}{I_{cl} + I_A} \text{ for } °C$$

$$= 0·061 \times \frac{T_S - T_O}{I_A + I_{cl}} \text{ for } °F$$

Since $I_A = I_{S.A.} - W$

$$H = 0·11 \times \frac{T_S - T_O}{I_{cl} + I_{S.A.} - W}$$

$$\therefore H(I_{cl} + I_{S.A.} - W) = 0·11(T_S - T_O)$$

$$\therefore H(I_{cl} + I_{S.A.}) = 0·11(T_S - T_O) + H.W.$$

$$= 0·11\left[T_S - \left(\frac{T_O - H.W.}{0·11}\right)\right]$$

$$\text{or } H = 0·11 \frac{T_S - \left(T_O - \frac{H.W.}{0·11}\right)}{I_{cl} + I_{S.A.}} \text{ for } °C$$

This means that the actual heat loss at temperature T_O with this wind is the same as it would be in still air at a temperature $\left(T_O - \dfrac{H.W.}{0 \cdot 11} \right)$. The 'thermal wind decrement' is therefore given by $\dfrac{H.W.}{0 \cdot 11}$°C or $\dfrac{H.W.}{0 \cdot 061}$ °F. We can now make a table of thermal wind decrements for different wind speeds, for a total heat loss of 1 Met. For heat losses of 2 Mets., 3 Mets., etc., the decrements in the table are simply to be multiplied by 1, 2, 3, etc.

THERMAL WIND DECREMENTS FOR TOTAL HEAT LOSS OF 1 MET.

Air Movement									
Ft./min.	20	35	50	75	120	210	425	1,050	4,500
Miles/hr	0·23	0·40	0·57	0·85	1·37	2·39	4·85	11·9	51·0
Thermal Wind Decrement									
°C	0	1·8	2·7	3·6	4·5	5·5	6·4	7·3	8·2
°F	0	3·3	4·9	6·6	7·4	8·2	11·1	13·2	14·8

The thermal wind decrement is thus simply the product, in appropriate units, of the total heat loss and the insulation-wind-decrement. Possibly this result may be astonishing, but a little thought shows that it is correct. To an object without heat production, or heat loss (such as an ordinary thermometer or an unheated house) which is in equilibrium with the environmental temperature, the wind makes no difference at all. In the case of a heated house, or the heated kata-thermometer, which is in a steady state rather than equilibrium, the wind speed has an effect. The effect is the larger the greater the heat loss or heat production. Thus the Englishman sitting in his unheated office is no colder when the wind blows outside (assuming the ideal case where the window frames did not leak) but those who live in heated houses know well that the furnace must be turned up when the wind blows harder or comfort will decrease.

Again, the indication from the table that the maximum thermal wind decrement amounts to about 8°C or 15°F may be surprising to some. Actually it corresponds with experience. Take the case of a man sitting in ordinary indoor clothing, 1 clo unit, at the comfortable temperature of 20°C (68°F) with low air movement. If the air movement is increased greatly he will be uncomfortably cool, the equivalent still-air temperature is now as low as 12°C (53°F), and in experiments on subjects lying in such a room with high air movement it was found that for comfort the temperature had to be raised to about 28°C (82°F). In the cold, it may appear that the wind makes much more difference than indicated by the table. For example, surely −18°C (0°F) with a high wind is equivalent to a still-air temperature lower than −25°C (−15°F)? Here again, we are thinking of the experience of cold on the un-protected parts of the body such as the face. This has a total insulation of only 1 clo in ideally still air. The heat loss at −18°C (0°F) will be then at least 5 Mets. even in still air, and for a steady state the rate at which heat is brought up to the face by the cir-culation, rather than 'heat production', would also have to be 5 Mets. For such a rate of heat loss the figure for the thermal wind decrement would be five times as great, i.e. 41°C (74°F), and the equivalent still-air temperature −59°C (−74°F). We would conclude that a very high wind at −18°C (0°F) would be in-tolerable to the unprotected face, a conclusion which agrees with experience.

Finally, it may seem that since, at the outset, the objection to the soundness of the wind-chill scale was that it took no account of the clothing worn, and that theoretically it was impossible to omit this, nothing has been achieved by introduction of the thermal wind decrement. Are not the effects of clothing insulation also ignored here? The answer is that the new way of estimating the effect of wind, as a decrement to the temperature, is sound, since after the equivalent still-air temperature is obtained the effects of the clothing insulation have then to be taken into account in estimating the thermal demand. A thermal wind decrement of the temperature by 10°C may be of little consequence to a heavily clothed body. On the other hand, the scale of wind-chill is sup-posed to be itself an index of thermal demand, not of equivalent temperature, and there is no way in which the amount of clothing insulation can be formally introduced.

The use of the equivalent still-air temperature has certain advantages of convenience. The graph of total insulation required (clothing plus air, Fig. 30) can be redrawn to show the clothing insulation required at a given equivalent still-air temperature (same as Fig. 30 except that the scale of insulation is reduced by 1 clo unit and the insulation of standard still-air and the abscissae are now equivalent still-air temperatures). The same chart then

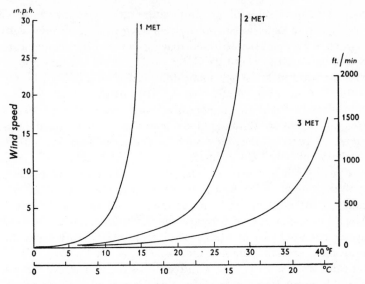

Fig. 32. The thermal wind-decrement to be subtracted from the thermometer reading to give the equivalent still-air temperature. Note that it depends on the metabolic activity of the man.

appears for all wind speeds. The multiple curves for different activities, however, are still necessary. Fig. 32 can be used to give the thermal wind decrement, instead of the table. It shows plainly how the decrement approaches a constant value for winds above 5 m.p.h., so that for practical purposes in most outdoor conditions we can assume that it is 5 to 8°C (10 to 15°F) for activity of 1 Met., and correspondingly more for greater activities.

Application to Meteorological Data

Conventionally, meteorological data as to temperature and wind speed have been given as mean values, for the day or month. For the climatic physiologist, mean values are not very useful. What he wants to know is for how much of the time the temperature will be at a given level, i.e. 'frequency data' are required rather than means. An example of how unrealistic the use of means, or of maximum values in meteorological data may be in predicting requirements, is that those charged with providing insulated shelters for the Arctic were told they must be adequate for temperatures down to $-48\cdot4$°C (-55°F) and wind velocities of 60 m.p.h. Actual frequency data would show that the temperature with the given velocity was below −35°C (−30°F) for less than 1 per cent of the time, and an enormous saving in fuel and cost of huts could be made by taking into account the frequency data. Again, though temperatures of −48·4°C (−55°F) had been recorded, and winds of 60 m.p.h., how often do these extreme conditions occur together? It is the opinion of most of those who have lived in the Arctic that when these extremely low temperatures are reached the wind velocity is seldom more than slight, and sound

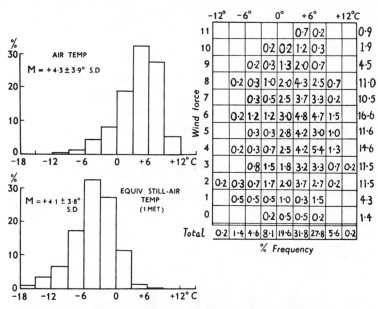

FIG. 33. Meteorological data for a square of the N. Atlantic (from work of R.N. meteorological branch) given in the form of frequency or contingency tables, and the application to it of the thermal wind decrement.

reasons from the physics of the atmosphere could be advanced to suggest this. Analysis of the correlation between wind velocities and temperatures has apparently not been made. Double frequency tables, or 'contingency tables', are required to show how often combinations of wind and temperature occur. Recently such tables have been prepared by the meteorological service of the Royal Navy, at the request of the R.N. Personnel Research Committee, for different areas of the North Atlantic Ocean, and give most valuable information. Fig. 33 shows such a double frequency table for one particular area. The histogram of the temperature, from the figures in this table, is shown. Below is the histogram of the equivalent still-air temperature, prepared by applying the thermal wind decrement to the temperature data. The mean still-air temperature is 8·4°C (15·1°F) lower than the mean temperature. The histograms appear to be close enough to normal distributions so the whole data, as regards thermal demand, for this geographical square could be usefully represented by the statistic-mean equivalent still-air temperature = 4·1 ± 3·8°C S.D. (39·4 ± 6·8°F S.D.). The standard deviation can be interpreted in the well-known manner. Equivalent still-air temperatures more than twice the S.D. below the mean, i.e. below −13·1°C (8·5°F) can be expected about 2 per cent of the time. Thus by the use of a unitary system of this kind for combining wind and temperature, a very great economy of expression of climatic data can be achieved. Also laboratory experiments, made in the cold chambers at low wind velocities, could determine the adequacy of clothing assemblies proposed for issue, and be interpreted in terms of field conditions with some confidence.

Realization that frequency rather than mean data is required has led Siple (2) to make complete 'thermal analyses' of American climates in these terms, for temperature, wind, solar radiation and humidity. The climate of a given place can be summed up graphically in a 'profile' (Fig. 34). Much more of this type of analysis

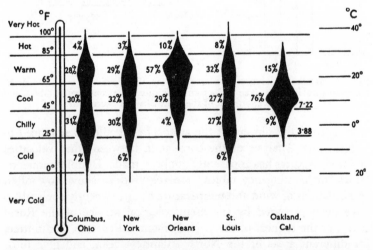

FIG. 34. Profiles of distribution of temperature for localities in the U.S. prepared by Siple. The width of the diagram indicates the percentage of the whole year the temperature is in the given range. (Siple. *Bull.Am.Inst.Architects*, 1949.)

needs to be done for the advancement of applied 'physiological human engineering'.

Solar Radiation

The layman is aware that when the sun shines it is 'much warmer', and our next task is to find a way in which the effect of the heat absorbed by the human body can be expressed, and if possible a 'thermal radiation increment' devised as a correction to the environmental 'shade temperature'. The heat received by the body in full sunshine can amount to two or three times the metabolic rate, so we would expect this increment to be quite large in such circumstances.

Again, we have to do some simple algebraical juggling with the fundamental equations, but here we must go back a step into the derivation of these equations, since the solar heat will be absorbed not on the body surface, but on the surface of the clothing. Fig. 35

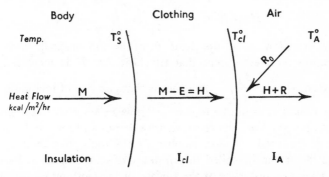

Fig. 35. Diagram to illustrate calculation of thermal radiation increment.

illustrates the elementary calculation required (3). The non-evaporative heat loss from the body surface is H cal/sq. m/hr, and for a steady thermal state this will be equal to $(M - E)$, the heat production minus the evaporative heat loss. This heat H will be transmitted through the clothing insulation I_{cl}, and must be lost from the surface of the clothing, to the environment, through the insulation of the air I_A. However, the surface of the clothing will be absorbing heat R cal/sq. m/hr from solar radiation, and in the steady state this amount in addition must be lost from the clothing surface.

I

The equation for the heat transfer from skin surface to clothing surface is therefore:

$$H = \frac{T_S - T_{cl}}{I_{cl}}$$

but for the heat loss from the clothing the equation is:

$$H + R = \frac{T_{cl} - T_A}{I_A}$$

these equations yield:

$$T_S - T_{cl} = H I_{cl}$$
$$T_{cl} - T_A = (H + R) I_A$$

adding:

$$T_S - T_A = H I_{cl} + (H + R) I_A$$

the right hand side could be rearranged to:

$$T_S - T_A = \left(H + R \times \frac{I_A}{I_{cl} + I_A} \right) \left(I_{cl} + I_A \right)$$

or:

$$H + R \times \frac{I_A}{I_{cl} + I_A} = \frac{T_S - T_A}{I_{cl} + I_A}$$

This equation is of the same form as the original, with no radiation received, except that the heat loss H is increased by the term $R \times \dfrac{I_A}{I_{cl} + I_A}$. This means that one way of including the solar radiation would be to consider it as an addition to the metabolic rate. We learn, however, that we do not add the whole radiation R absorbed to the metabolism but R multiplied by a factor. This factor can be called the *efficiency of solar radiation*. Part of the solar heat absorbed at the surface of the clothing can be considered as lost to the environment, the rest as increasing the heat load of the body.

Efficiency of solar radiation =

$$\frac{\text{Insulation outside point of absorption}}{\text{Total insulation}}$$

For a man clothed for the Arctic, with a total of 5 clo units of insulation, and average insulation of the air, say 0·5 clo units, this efficiency is quite low (10 per cent). We can see why the desert Arab wears his burnous rather than the smallest possible amount of clothing insulation, for by absorbing the solar radiation beyond the body he may effectively decrease the heat load, even though his thermal insulation be increased. In addition, of course, the white clothing absorbs far less of the incident radiation than would his skin (see later). It is curious that some desert Bedouin live in black tents and wear black clothes: this may be an illustration of the dominance of cultural prejudices over functional considerations.

This, however, is not the method of accounting for the solar heat load we have set out to devise, though it may help our understanding of the problem to have used this concept.

The equation may also be written:

$$T_S - T_A = H\left(I_{cl} + I_A\right) + R.I_A$$

$$\text{or } H = \frac{T_S - \left(T_A + R.I_A\right)}{I_{cl} + I_A}$$

The equation shows that the radiation increment to the temperature is very simple, i.e. the product $R.I_A$, of the radiation absorbed and the insulation of the air. Working in Met. units for R, and clo units for I_A,

$$\text{Radiation increment in } ^\circ\text{C} = \frac{R.I_A}{0\cdot 11} = 9\cdot 1\ R.I_A$$

$$\text{Radiation increment in } ^\circ\text{F} = \frac{R.I_A}{0\cdot 061} = 16\cdot 1\ R.I_A$$

The Amount of Solar Radiation Absorbed

To calculate this is a very complicated matter, since so many factors are involved. These depend on the posture of the man, i.e. the cross-sectional area presented by the body normal to the sun's rays, the reflecting power for radiation of the surface of the clothing, the altitude of the sun, the absorption by moisture and dust in the air and by clouds, the scattered radiation from all directions, the diffuse reflection from the ground, and so on. All these have been carefully analysed by Blum (4 and 5), who finds that in spite of all these variables it is possible to arrive at a useful average figure, for the radiation incident on the body, on a cloudless day, for the daylight hours. Some of the important factors vary in opposite directions; for example, when the sun is low the intensity of the radiation decreases because of the greater absorption by the atmosphere, but the cross-sectional area offered by an erect man increases. For a clear sky, the mean radiation incident on a man amounts to about 230 kcal/sq. m/hr, or about 4·6 Mets.

The reflecting power of ordinary clothing is considerable: it has been measured for a number of fabrics. Black clothing may reflect 12 per cent and absorb 88 per cent of solar radiation, white clothing on the other hand, will reflect 80 per cent and absorb only 20 per cent. For military camouflage khaki or 'olive-drab' clothing, the reflecting power is about 43 per cent, and absorption 57 per cent. A considerable part of the total load on the body comes from reflection and re-radiation from the terrain, and the absorbing power of clothing for this radiation, of much longer wavelengths than directly from the sun, will be greater for all colours.

For military clothing the mean solar load, from an unclouded sky, may amount to 130 kcal/sq. m/hr. This is about $2\frac{1}{2}$ Mets., equal to the heat production of a man in moderate exercise. Obviously we cannot neglect this factor in the heat balance of man in the cold during the daylight hours.

How are we to arrive at a figure for R, the solar heat load, in actual conditions? In well equipped meteorological stations the solar heat incident on a black horizontal surface is recorded continuously, but even this fluctuates from minute to minute with clouds so that it is difficult to strike an average. Fortunately meteorologists have a system of describing the degree of cloudiness in tenths, e.g. 3/10 cloudy, and the radiation can be calculated from this. Of course, the absorption depends on the thickness and nature of clouds, so that a complete overcast may reduce the radiation to as little as 3 per cent for very heavy cloud or to 20 per cent for very thin cloud. For average cloud, 10 per cent for complete overcast could be assumed, and therefore:

$$R = R_O(1 - 0.9x)$$

where R is the radiation energy, R_O that above the clouds, and x is the 'average cloudiness'. There is a good deal of evidence that in northern latitude the cloud cover is thinner, and a complete overcast may reduce the radiation only to 40 per cent in Arctic regions, so that predictions of the thermal radiation increment made on the above equation are too low for such regions.

The final equation for R, in Met. units, is therefore:

$$R = 4 \cdot 6 \, (1 - 0 \cdot 9x) \times \frac{a}{100} \text{ Mets.}$$

where x is the average cloudiness, and a the per cent absorbing power of the clothing, taken as 88 per cent for black clothing, 57 per cent for military khaki, and 20 per cent for white clothing.

The thermal radiation increment is given by multiplying R by the insulation of the air I_A appropriate to the prevailing wind, as already shown. The final equation becomes:

$$\text{Radiation increment } T_R = 9{\cdot}1 \times 4{\cdot}6\,(1 - 0{\cdot}9x) \times \frac{a}{100} \times I_A$$

$$= 0{\cdot}42\,(1 - 0{\cdot}9x)\,a.I_A \text{ in } °C$$

$$= 0{\cdot}76\,(1 - 0{\cdot}9x)\,a.I_A \text{ in } °F$$

The equation above lends itself to the construction of a chart from which the values of the thermal radiation increment can be

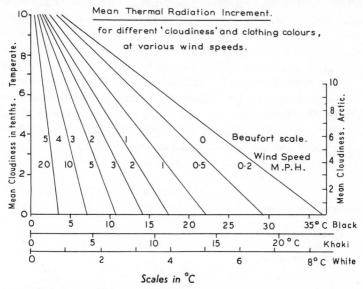

FIG. 36. Chart for estimating the thermal radiation increment to be added to the temperature to give the equivalent shade-temperature.

read, for various degrees of cloudiness, white, black or khaki clothing, and wind speeds (Fig. 36).

The chart shows that radiation may increase the effective temperature, for military clothing, by as much as 19·4°C (35°F) if the wind speed is low and the sky unclouded, and that even in average conditions the increment may be several degrees. Blum made the complete calculations from meteorological data for the months of July and January in the north temperate zone. He found the average was an increment of about 5°C (9°F) at 40,000 ft. for July, and 4·45°C (8°F), at 40,000 ft. for January. Similar calculations for January and February at Fort Churchill, Manitoba, showed that over a period of six weeks the increment in daylight hours averaged 4·9°C (8·8°F), and days on which it was over 8·3°C (15°F) were not infrequent. With multiple variables involved, the chart should of course be used merely to give an indication of the magnitude of the thermal radiation increment from the meteorological data that are commonly available.

Validation of the Theory

A rough check on the agreement of the predicted increment and actual experimental results is possible from the work of Robinson *et al.* (6), on three subjects sitting in the sun, wearing only shorts. Here the sweating was so great that the heat loss by evaporation was considerably larger than the heat production, and the difference gives an estimate of the heat loss by radiation and convection. This is negative because of the large amount of heat being received by the body from solar radiation. Dividing by the insulation of the air, the effective thermal gradient from air to skin is obtained, and comparing this with the actual gradient, skin temperature to air temperature, the thermal radiation increment can be estimated. The calculations are given in the table below:

		M.D.	*E.T.*	*J.P.*
Surface area, sq. m		1·70	1·96	1·71
O₂ consumption, l./min.		0·30	0·31	0·30
(1) O₂ consumption, cal/hr		86	90	86
Evaporation, g/hr		521	555	529
(2) Evaporation, cal/hr		313	333	317
Difference (1) — (2), heat by convection and radiation from body cal/hr		−227	−243	−231
(3) cal/sq. m/hr		−133	−124	−135
(4) Average skin temp.	°C (°F)	36·0(96·8)	35·6(96·1)	34·6(94·3)
(5) Air temp.	°C (°F)	34·8(94·6)	34·4(93·9)	34·8(94·6)
(6) Difference (4) — (5) Thermal gradient	°C (°F)	1·2(2·2)	1·2(2·2)	−0·2(−0·3)
Wind velocity, miles/hr		3·5	3·4	3·5
(7) Insulation of the air, clo units		0·34	0·34	0·34

(8) Effective thermal gradient $\dfrac{(3)\times(7)}{5\cdot55}$ °C $-8\cdot2$ $-8\cdot3$ $-8\cdot3$

$\dfrac{(3)\times(7)}{3\cdot09}$ °F $-14\cdot8$ $-15\cdot0$ $-15\cdot0$

Radiation increments
(6)$-$(8)   °C (°F) $9\cdot4(17\cdot0)$ $9\cdot5(17\cdot2)$ $8\cdot5(15\cdot3)$

The chart would indicate that the increment should be about 8·3°C (15°F) (human skin is similar in absorbing power to khaki clothing and the sky was unclouded). The prediction is at least correct as to the order of magnitude.

Greenhouse Effect

If it were possible to have translucent clothing which permitted most of the solar radiation to penetrate and be absorbed close to the skin, while thermal insulation was retained, quite fantastic increments could be achieved. The formula in this case for the radiation increment is $R(I_A + I_{cl})$, instead of $R.I_A$. A translucent suit of 3 clo units insulation would give an increment of 54·4°C (130°F) for unclouded sky, even in a high wind ($I_A = 0\cdot1$ clo). An inactive man could be quite comfortable at -18°C (0°F) when the sun shone, though very cold when it was obscured. The possible application of the 'greenhouse' principle to hand-gear merits some attention.

The Still-Shade-Temperature

If we subtract the thermal wind decrement, and add the radiation increment, we arrive at the equivalent 'still-shade-temperature'. It is of interest to note how the wind speed enters into both of these corrections to the temperature, and in such a way that it has a double effect in increasing the 'coldness' of the environment; not only does the wind decrement increase with greater wind speed, but the radiation increment greatly decreases. Thus in outside conditions of sunshine, the wind speed becomes of the greatest importance in determining the comfort and endurance of man. If the wind speed is 5 m.p.h., the thermal decrement for a resting man is about equal to the radiation increment for full sunshine, so that if the wind is greater than 5 m.p.h. it will be 'colder' to stand outside in the sun than inside a shelter at the same temperature. For a man of activity 2 Mets., the critical wind speed is less than this, about 3 m.p.h. and for activity 3 Mets., only about $1\frac{1}{2}$ m.p.h.

When the sky is cloudy, of course, the still-shade-temperature will be less than the dry bulb temperature for the lowest air movements. All this emphasizes the great importance of shelter from the wind in living in the cold.

Other Attempts at Unitary Systems

A classical attempt at expressing a further environmental factor affecting human thermal comfort, namely the relative humidity, as a correction to the temperature is the deservedly famous 'effective temperature' scale of Houghton and Yaglou (7 and 8). This was devised from 'instantaneous' thermal impressions of subjects while passing back and forth from one conditioned room, a given temperature and very low humidity, to another, at lower temperature and higher humidity. One of its authors has discussed its shortcomings as an index of the continuous thermal demand on the body of the environment (9) because of its dependence on these initial impressions. Effective temperature has been shown, even in the range of ordinary comfort temperatures in air-conditioning problems, to overestimate the effect of humidity, while at very high temperatures it greatly underestimates the effects of humidity (10). It is clear that for the problems of cold, this scale can have no validity for including the humidity with the effects of wind and radiation we have already considered, for the effective temperature scale predicts that higher humidities increase 'warmth' of the environment in the range of temperatures it covers, and gives us no indication that this will be reversed at lower temperatures. In view of the complicated effects of humidity in clothing already discussed, which we are far from understanding fully, it appears that for the present we cannot devise any unitary scheme that includes humidity. Bedford (11) has devised a corrected effective temperature scale which takes into account radiation.

Another unitary index is the 'Operative Temperature' of Winslow, Herrington and Gagge (12). This takes into account velocity of air movement and radiation, but to find the operative temperature, the 'mean radiant temperature' from some type of globe thermometer is required, as well as the skin temperature. It therefore cannot be estimated from the ordinary meteorological data as can the still-shade-temperature discussed above. Also the experiments on which the equations are based were made on unclothed subjects in a booth with highly polished reflecting walls, with radiation from heaters, and it would be doubtful if the results have any application to the general problem of the natural outside environment.

An attempt to combine most of the relevant factors in a single index is the 'Thermal Acceptance Ratio' of Ionides, Plummer and Siple (13). This defines the maximum skin temperature and skin vapour pressure that are considered tolerable for man, and from these calculates the heat loss to the environment. The acceptance ratio is the ratio of this to the heat production. The effect of wind speed appears not to be included, nor that of clothing, and in any case this type of analysis would appear to be useful only for hot environments.

Conclusion

It is concluded that the variables other than temperature that affect the heat loss of a human body, namely the wind speed and the radiation from the sun and terrain, can be included in a unitary scheme as corrections to the environmental temperature. However, the metabolic rate of the man persists as a dominant factor in determining his comfort, and no inclusion of this as a correction factor is possible. We must continue to think in terms of sets of requirements that are different for each level of activity, and have separate standards for the resting man, for the man in light exercise, the marching soldier and the man in violent combat or in occupations involving heavy work.

REFERENCES

1. SIPLE, P. A. and PASSEL, C. F. Dry Atmospheric Cooling in Sub-freezing Temperatures. *Proc. Am. Philos. Soc.*, **89,** 177, 1945.
   See also: COURT, A., Wind Chill. *Bull. Am. Met. Sc.*, **29,** 487, 1948.
2. SIPLE, P. A. American Climates, *Bull. Am. Inst. Architects*, Sept. 1949.
3. BURTON, A. C. An Analysis of the Physiological Effects of Clothing in Hot Environments. Associate Committee on Aviation Medical Research. No. C. 2754, 1944.
4. BLUM, H. F. The Solar Heat Load, its Relationship to the Total Heat Load, and its Relative Importance in the Design of Clothing. Naval Med. Research Institute Report, May, 1944.
5. BLUM, H. F. Calculation of Solar Radiation Intensity and Solar Heat Load, on Man at the Earth's Surface and Aloft. Memorandum Report No. T.S.E.A.A. 695, Feb. 1946.
   (Army Air Forces Technical Service Command, Engineering Div., Aero-Medical Laboratory.)
6. ROBINSON, S. and TURRELL, E. S. Studies of the Physiological Effects of Solar Radiation. Int. Report No. 11, from the Dept. of Physiology, Indiana University Medical School, May 14, 1944.

7. HOUGHTEN, F. C. and YAGLOU, C. P. Determining Lines of Equal 'Comfort'. *Tr. Am. Soc. Heat. Vent. Engr.*, **29**, 163, 1923.
8. YAGLOU, C. P. and MILLER, W. E. Effective Temperature with Clothing. *Tr. Am. Soc. Heat. Vent. Engr.*, **31**, 89, 1925.
9. YAGLOU, C. P. 'Indices of Comfort'. Physiology of Heat Regulation, p.277, ed. Newbergh, 1949.
10. ROBINSON, S. and GERKING, S. D. Thermal Balance of Men Working in Extreme Heat. *Am. J. Physiol.*, **149**, 476, 1947.
11. BEFORD, T. 'Basic Principles of Ventilation and Heating'. Lewis, London, 1948.
12. GAGGE, A. P. Standard Operative Temperature, a Single Measure of the Combined Effect of Radiant Temperature, of Ambient Air Temperature and of Air Movement on the Human Body. In Temperature: Its Measurement and Control in Science and Industry. American Institute of Physics. Reinhold Publishing Co., New York, 1941.
13. IONIDES, MARGARET, PLUMMER, J. and SIPLE, P. A. Thermal Acceptance Ratio. Climatology and Environmental Protection Section. Military Planning Division, Office of Quartermaster General Int. Report No. 1, Sept. 17, 1945.

CHAPTER 8

VASCULAR REACTIONS TO COLD

The mechanisms involved in the regulation of body temperature include a control of the peripheral circulation. Heat loss from the surface of the body is modified by the rate of blood flow through the skin and superficial tissues. When the blood flow is reduced there will be, in a cold environment, a much steeper gradient of temperature from the deep tissues to the skin, and heat loss is reduced because of a low skin temperature.

Effect of Local Temperature Changes

The effect of local changes of temperature on the circulation has been extensively investigated. Barcroft and Edholm (1) showed that the blood flow in the forearm diminished with reduction of the temperature of the water, in which the arm was immersed. However the temperature of the blood flowing through the vessels was also reduced, and therefore the viscosity of the blood increased. Such an increased viscosity would account in part for the diminished blood flow, but there was also vasoconstriction contributing to the decreased flow (Fig. 37).

The forearm blood flow averaged 3 ml./100 ml. forearm/min., and the deep muscle temperature was approximately 37°C (98·6°F) when the arm was immersed in water at 34°C (93°F). The blood flow was reduced to 0·75 ml./100 ml. forearm/min. at a water bath temperature of 15°C (59°F), where the deep muscle temperature was only 18·5°C (65·3°F). The temperature of the blood flowing in the arterioles may be assumed to be approximately equal to the deep muscle temperature, so in these two conditions the blood temperature would be 37°C (98·6°F) in a water bath at 34°C (93°F) and 18·5°C (65·3°F) in a water bath at 15°C (59°F). The effect of temperature on the viscosity of blood is similar to that on the viscosity of water. The ratio of viscosity at 18·5°C (65·3°F) to that at 37°C (98·6°F) is $\dfrac{105}{70}$.

Since the blood pressure was constant and assuming that there had been no vasoconstriction the blood flow would be reduced from 3 ml. to 2 ml./100 ml. forearm/min. In fact, the flow fell

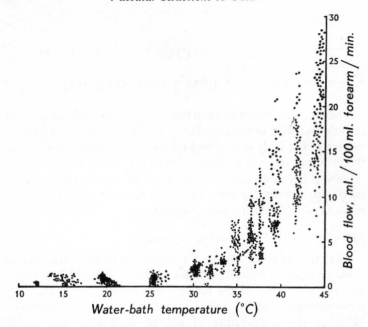

Fig. 37. Forearm blood flow plotted against water bath temperature. The forearm and hand was kept in water at a particular temperature for two hours. The blood flow declines with decrease of temperature. (Redrawn from Barcroft and Edholm, *J. Physiol.* **102**, 5, 1943.)

from 3 ml. to 0·75 ml., so there was a significant vasoconstriction in addition to the viscosity effect. This vasoconstriction was shown to include the muscle vessels as well as the cutaneous vessels. Similar effects were also observed in subjects who had been sympathectomized, and therefore the effect of cold is in part due to a direct effect on the blood vessels.

Cold Vasodilatation

When very low water bath temperatures are employed the opposite effect, i.e. increased blood flow in the hands and fingers, has been observed. This was first studied by Lewis (2), who measured the skin temperature of fingers immersed in ice water. The temperature fell rapidly almost to 0°C (32°F) but 10 to 15 minutes later the skin temperature rose by some 5° to 6°C (9° to 11°F). Characteristically with continued immersion the finger temperature fluctuated slowly between 0°C (32°F) and 5° to 6°C (41° to 43°F). Lewis concluded that the rise of temperature was due to increased blood flow and he termed the fluctuations in skin temperature the 'hunting' phenomenon (Fig. 38).

Grant and Bland (3) made further studies, including microscopic observation of the rabbit's ear, and produced convincing evidence that the 'cold dilatation' was due to an increased flow through the arteriovenous anastomoses. It was also shown by Lewis that the effect persisted after sympathectomy and peripheral nerve section. He could not produce a cold vasodilatation, when the peripheral nerve had degenerated and he concluded that an axon reflex was

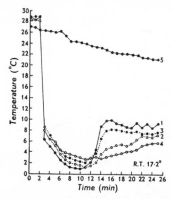

Fig. 38. Skin temperature recorded with thermocouples on right index finger (1, 2, 3, 4) and temperature of dorsal surface, base of nail of left index finger. At time 2 minutes, right index finger immersed in crushed ice. Note the abrupt fall of skin temperature of this finger, followed by a rise of temperature after 10 minutes, which is more marked in the distal than in the middle phalanx. 1. Tip of right index finger. 2. Middle of ventral surface, middle phalanx. 3. Dorsal surface, base of nail. 4. Dorsal surface, middle phalanx. 5. Dorsal surface, base of nail, left index finger (control). Room temperature 17·2°C. (Redrawn from Grant and Bland, *Heart*, **15**, 385, 1931.)

involved. Kramer and Schulze (4) also considered that the cold vasodilatation was due to an axon reflex as they succeeded in preventing the cold vasodilatation by blocking the finger tip with novocaine. This observation was not confirmed by Greenfield *et al.* (8) (*vide infra*).

Aschoff (5a to g), in a series of papers, described the factors influencing cold vasodilatation. He used a constant flow calorimeter and measured the heat output of the whole hand. Using water at 10° to 13°C (50° to 55°F), the heat output was at the lowest values obtained, being 25 to 30 calories/minute. When the water temperature was reduced below 6°C (43°F), the heat output increased abruptly after 10 or 15 minutes, up to values of 150 to 200 calories/minute. Aschoff demonstrated fluctuation in heat output during cold vasodilatation, corresponding to the fluctuation of skin temperature, i.e. the 'hunting' phenomenon. Intracutaneous temperatures in the finger and hand at various sites showed a steep

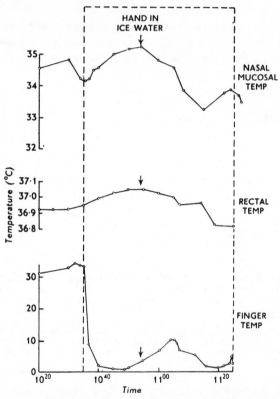

FIG. 39. Changes in nasal mucosal, rectal and finger temperature, during immersion of hand in ice water. The time of immersion is shown by the interval between the dotted lines. There is an initial fall in mucosal temperature, which rises again as rectal temperature goes up, owing to the generalized vasoconstriction of the peripheral vessels when the hand is put in ice water. The arrows indicate the onset of cold vasodilatation. As heat loss increases from the hand rectal temperature and nasal mucosal temperature fall. (Redrawn from Aschoff, *Pflügers Arch.*, **248**, 436, 1944.)

fall on immersion in ice water and then a rise of as much as 10°C (18°F). During the phase of dilatation (Fig. 39), Aschoff found that the temperature increase was greatest in the terminal phalanx and developed in advance of the temperature rise in the proximal phalanx, which in turn showed a higher and earlier rise than that in the hand. This finding, indicating a marked gradient of blood flow from finger tip proximally, has since been confirmed by Greenfield *et al.* (8c). In Aschoff's experiments the heat output of the finger per unit surface was five times greater than that of the whole hand. This observation is reasonably explained by the distribution of the arteriovenous anastomoses $(A - V)$. Clara (6) estimated that there are 500 AV/cm² in the nail bed, 236/cm² in the finger tip, and only 93/cm² in the proximal phalanx. There are considerably fewer in the hand. Aschoff suggested that cold paralysis of the vasoconstrictor fibres might be partly responsible for the 'hunting' phenomenon. During paralysis, the vessels dilate, the blood flow increases and the finger warms up. The paralysis passes off and vasoconstriction follows with a fall in blood flow and temperature, and then cold paralysis once more develops. It should be made clear that the 'hunting' phenomenon refers to the alternating periods of increased and decreased blood flow commonly observed in the finger on immersion in cold water. It is not unusual to observe persistent cold vasodilatation without any periods of vasoconstriction, i.e. the 'hunting' phenomenon is absent.

Spealman (7) made direct measurements of the blood flow in the hand at different water bath temperatures, using a plethysmograph. He showed that the blood flow decreased with decline of the water bath temperature until this reached 10°C (50°F). When the temperature of the water bath was further reduced there was a striking increase in the hand blood flow, which was frequently as large as at the relatively high temperature of 35°C (95°F). The local effects of temperature were considerably influenced by the general thermal state of the body, a finding which has since been abundantly confirmed. If the subject was comfortably warm, the effects were as already described, but when the subject was uncomfortably hot the blood flow in the hand remained high even when it was in water at 10°C (50°F), so the local vasoconstriction effect of cold water at this temperature was suppressed. On the other hand, if the subject was chilled, the increased blood flow at low water temperatures was reduced, although some cold vasodilatation was still observed.

The phenomenon of the vasodilator effect of cold has been extensively investigated by Greenfield and his colleagues (8a to f), using a calorimeter to estimate blood flow in the finger, the hand, the toes and the foot. It was found that plethysmography was not a suitable technique to measure blood flow at very low temperatures.

During the phase of cold vasodilatation the inflow curves recorded, when a collecting pressure was applied, flattened after only one or two pulse beats, so it was not possible to make an accurate determination. This difficulty, which surprisingly was not mentioned by Spealman, appears to be due to a great reduction of the venous reservoir. There may be a direct effect of cold constricting the veins or so increasing their tone that the venous pressure rises very rapidly when venous occlusion is applied. The effect

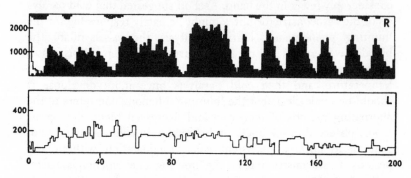

FIG. 40. The heat loss from the left index finger immersed in ice water, shown in the top panel, compared with the heat loss from the right index finger kept in water at 30°C. Time, on the abscissa, is given in minutes. Heat elimination is in calories, note the difference in scales between the two panels. Pain is indicated by the marks below the top line of the upper panel, the thicker the mark the greater the pain. Heat elimination is directly related to blood flow and the variations shown in the top panel indicate alternating vasodilatation and vasoconstriction. (From Greenfield, Shepherd and Whelan, *Irish J. Med. Sci.*, 415–419, September, 1951).

of low temperature on veins is a subject which deserves future study. Greenfield *et al.* (8a, b) found that when finger or hand is immersed in ice water there is an immediate constriction which may be so intense as to occlude the local circulation completely. After approximately five minutes there is a sudden very large increase of blood flow, which may be sustained, but in most subjects periods of vasoconstriction and vasodilatation are observed (Fig. 40). The thermal state of the body as a whole affects the local vasodilatation, as Spealman found. The cold vasodilatation is greatly increased when the subject is hot, and reduced, but not abolished, when the subject is cold. Lewis (2) noted that cooling of the forearm alone suppressed cold vasodilatation in the finger. Greenfield *et al.* (8d) also found, confirming Lewis, that the local vasodilatation is not dependent on the vasomotor system, as it is obtained in sympathectomized subjects. It will be recalled that previous workers considered that an axon reflex was involved in the mechanism of cold vasodilatation, but present evidence throws doubt on this view; it has been shown (8d, f, 9) that in subjects with complete section of nerves to finger and subsequent degeneration, cold vasodilatation could still be obtained, although the response was considerably modified. The degree of dilatation was usually less in the denervated than in the control finger, and appeared gradually to increase during regeneration. If the denervated and control fingers were first immersed in water at 15°C (59°F) for 20 minutes, and then transferred to ice water, there was the usual response in the normal finger but the cold vasodilatation in the denervated finger was markedly reduced. On the other hand, if the fingers were warmed in water at 45°C (113°F) before immersion in ice water, the cold vasodilatation might be even larger in the denervated than in the normal fingers. Local preheating or precooling of the normal finger had no effect on the size of the response in ice water but considerably effected the response in a denervated finger. The degree and frequency of 'hunting' did not appear to be modified by denervation.

In normal subjects the 'hunting' phenomenon is most variable and it is difficult to predict the frequency of the waves of constriction. It is possible that the frequency increases with length of immersion, so there are more periods of constriction the longer the finger is immersed. 'Hunting' comes on at different times in the two hands and in fingers of one hand, although it may be so nearly simultaneous as to suggest central control. However if two fingers are immersed in ice water, the second going in five minutes after the first, it becomes clear that the time of appearance of cold vasodilatation is related only to the time of immersion of that finger, and subsequent periods of 'hunting' are not necessarily in step. That there may be frequently synchronism of 'hunting' is suggested by Aschoff (5g) as he found that systolic blood pressure fluctuates with 'hunting'.

There are seasonal variations in the degree of cold vasodilatation and the Lewis reaction is most easily provoked in July and August, according to Kramer and Schulze (4). They were also able to detect a periodic vasodilatation in cold air at temperature of 15° to −15°C (59° to 5°F). Blair (10) has demonstrated rises of skin temperature of the fingers in subjects exposed to low air temperatures,

K

but the reaction appears to be more variable in cold air than in cold water. Miller (11) observed phasic fluctuations in the skin temperature of fingers and toes during exposure, and Blaisdell (12) has recently reported detailed studies of cold vasodilatation in fingers exposed to cold air. This worker measured skin temperature and the amplitude of pulsation in the finger with a plethysmograph as an index of blood flow. His results show that on exposure to air at 0°C (32°F) the finger temperature gradually falls and then there are successive waves of warming and cooling associated with increase and decrease of the volume of finger pulsation. Blaisdell was able to obtain these results consistently, and the difficulty that previous workers have had in demonstrating vasodilatation in cold air may have been due to the conditions of their experiment. It is especially interesting that Blaisdell obtained cyclic variation even when his subjects were cold, and the height of the skin temperature rise did not appear to be affected by the thermal state of the body. However 'the degree of local cooling necessary to evoke the reaction is less the warmer the body'. Cold vasodilatation can be demonstrated in the toes and also in the skin of the nose and face (8e). The heat loss from the cold vasodilated hands immersed in ice water is very considerable, and can amount to as much as 40 kcal/hr. It is not surprising that a considerable fall in rectal temperature occurs when hands alone are maintained in ice water. The temperature of the finger tissues has been calculated to be as high as 30°C (86°F) during the height of vasodilatation in ice water, so there is a very steep gradient of temperature within the finger.

There is still no satisfactory explanation of the mechanism of cold vasodilatation. The work of Greenfield *et al.* strongly suggests that nervous pathways are not essential, so a humoral basis appears likely unless there may be a biophysical explanation. As Lewis and others have pointed out, whatever humoral substance is involved must be very stable. Whelan (9) has shown that histamine is probably not the humoral agent. It is possible that the responsible substance is only produced in the superficial layers of the skin, and is not washed away rapidly, as the increased blood flow is in the arteriovenous anastomosis with little flow in the capillaries.

In summary, an increased blood flow can be demonstrated in the extremities on immersion in water at temperatures between 0°C (32°F) and 8°C (46·4°F) or on exposure to air between 0°C

(32°F) and 15°C (59°F), in all cases following an initial vasoconstriction. The dilatation is usually but not necessarily periodic,
and can be demonstrated after complete denervation. Many problems remain to be answered. What is the exciting agent? For how

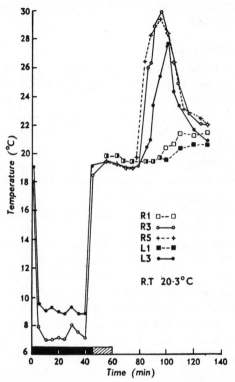

Fig. 41. Skin temperature of fingers immersed in water at 7°C (R1, R3, R5) and
at 9°C (L1, L3), from 0 to 44 minutes. Both hands then removed and put in
water at 19·4°C from 44 to 59 minutes. Hands taken out, and exposed to air
(Room T° 20·3°C). In 2/3 fingers previously in water at 7°C there is a marked
after dilatation in air, which is also shown by 1/2 fingers which were in water at
9°C. (Redrawn from Wolff and Pochin, *Clin. Sc.*, **8**, 145, 1949.)

long can the dilatation be maintained? So far the longest recorded
experiment is 3 hours. What is the relationship of this phenomenon to immersion foot and frostbite? How far is it modified by
local acclimatization of the hands to cold? Is the seasonal difference related to acclimatization?

On removal of the fingers from the ice water there is a further
vasodilator reaction (13) (Fig. 41). The size of the reaction is
related to the temperature of the bath: the colder the bath the

greater the effect. No significant after-reaction is observed except with bath temperatures of 10°C (50°F) or colder, and the after-reaction can be delayed by transferring the hands from the ice water to water at 25°C (77°F). The hands can be kept at this water temperature up to 2 hours, and the after-reaction still be obtained on removal.

Pain Produced by Cold

Wolf and Hardy (14) investigated the pain produced by immersion of a finger in cold water. At 18°C (64·4°F) fleeting deep ache developed 60 seconds after immersion, but at lower temperatures pain came on sooner and was more intense, the intensity at the peak of pain being directly related to temperature of the water. No spatial summation was noted: the pain caused by immersion of one finger was as severe as immersion of the whole hand. Whelan (9), however, considered that there was slight summation. Similar pain was experienced, according to Wolf and Hardy, on cooling the face, tongue, vertex and scrotum, but not in the lobe of the ear or the glans penis. These workers also showed that the blood pressure effect and the pain experienced were closely related. When the hands are immersed in ice water there is usually a sharp rise of blood pressure, and this forms the basis of the 'cold-pressor' test of actual or potential hypertensives. On prolonged immersion of the hands in cold water the systolic blood pressure may show rises and falls corresponding to the 'hunting' reaction (5g). Each phase of vasoconstriction is accompanied by pain (Fig. 10). Blaisdell (15) gives a detailed account of the changes in sensation in the hands on exposure to cold air and cold water. It is a striking experience to immerse the fingers in ice water. The initial pain comes on at first gradually and builds up to an almost intolerable intensity, then fades, as the cold vasodilatation develops. The subject may wonder if the water bath temperature has been raised as his finger feels comfortably warm. Then as vasoconstriction comes on pain once more develops although usually not to the same extent as initially, and as vasodilatation re-develops the finger will again feel warm and comfortable. Pain is therefore closely related to the level of blood flow.

Effect of General Thermal State on the Peripheral Circulation

It has already been mentioned that the effect of local cooling on hand blood flow can be greatly modified by warming or cooling the body as a whole. Bader and Mead (16, a, b) have further examined this effect and have shown that the finger blood flow did not decrease even if in ice water, provided the subject remained in an environment at 32°C (89·6°F). When the subject was cooled by sitting in a cold chamber at 0°C (32°F) and the hands

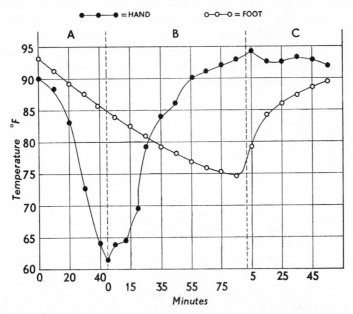

FIG. 42. Effect of rewarming the body on hand and foot temperatures. The subject sat in a chamber at −18°C (0°F), wearing ventilated clothing, light gloves and heavy boots. No heat was supplied during the first 45 minutes (A) and both hand and foot temperatures fell. As the insulation of the boots is greater than that of the gloves, the foot cools more slowly. For the next 95 minutes, heat was supplied, to the body only, in the ventilated clothing, reducing the over-all heat loss to 15 kcal/hr (B). Hand temperature rises rapidly, but the foot continues to cool. When the heat supplied is raised (C) so that the heat loss is only 6 kcal/hr, foot temperature also rises. (Redrawn from Rapaport *et al.*, *J. Appl. Physiol.*, **2**, 61, 1949.)

immersed in warm water the blood flow remained low. Rapaport *et al.* (17), using ventilated clothing with which any desired amount of heat could be supplied to the body, showed that if heat balance was maintained bare hands would remain warm even if exposed to an air temperature of −35°C (−31°F). However, if the heat supplied was reduced by even 15 per cent hands and feet cooled rapidly (Fig. 42). The practical importance of the dependence of the local circulation on general body temperature is obvious in clothing design (18a, b), and (19). It was shown that, if the body is cooling, electrical heating for the hands alone is useless and may be dangerous: it is more important to supply heat to the body than to the extremities. There are physiological limits to this argument, as shown by the fact that indirect vasodilatation in the forearm in response to heating the legs cannot be demonstrated unless the forearm be kept in a moderately warm bath. At 30°C (86°F), the blood flow in the forearm is reduced and there is either a slight or no response to leg heating; but if the arm is kept in water at 34°C (93°F), indirect vasodilatation can easily be demonstrated (Fig. 43). Local temperature effects and

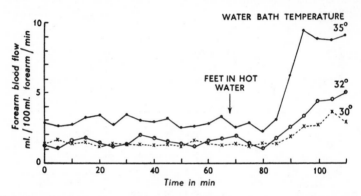

FIG. 43. Effect of local temperature on vascular response to heating the legs. Forearm blood flow was measured with the arm in water at 35°, 32°, and 30°C. At the arrow the feet and legs were put into water at 43·5–44°C. The size at the subsequent vasodilatation in the forearm is affected by the temperature of the water bath. (Redrawn from Barcroft and Edholm, *J. Physiol.*, **104**, 360, 1946.)

vasomotor effects tend to balance each other. The vasomotor control of the hand circulation and particularly of the digital vessels is more complete than that of muscle vessels, which may therefore be more affected by local temperature changes (20). The extent of indirect peripheral vasodilatation produced by heating various regions of the body, depends not only on the local and general environmental temperatures but also on the particular area that is heated. Bader and Macht (21) found that heating the face was far more effective than heating the chest or legs in raising skin temperature and blood flow of the hand. The subjects were seated in a room kept at 15°C (59°F), and the skin temperature of the hands fell to 16° to 20°C (61° to 68°F). Heating the chest with infra-red heat so raising the heated skin temperature to 42° to

44°C (107·6° to 111°F) did not increase hand blood flow or skin temperature, but heating the face, although the area of skin which was warmed was smaller, raised the skin temperature on the hands some 10°C (18°F) and the blood flow increased approximately fourfold.

Burton and Taylor (22) have demonstrated the effect of body thermal state on the peripheral circulation by recording the finger volume pulsation and the frequency of vasoconstriction. They showed that there is a regular series of vasoconstrictions in the finger and their frequency is related to the temperature of the

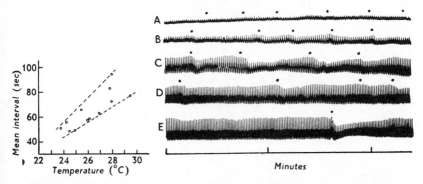

FIG. 44. Right-hand side; record of fingers volume-pulse in the same subject when he was cold, A, to very warm, E, showing the rhythm of periodic vaso-constrictions. The dots over the record show the periodic cardiac accelerations indicated by measurement of the intervals between systolic upstrokes. Left-hand side; modification of the mean interval between vasoconstrictions in the finger by environmental temperature. (From Burton and Taylor, *Am. J. Physiol.*, **129**, 565, 1940.)

environment. If the subject is uncomfortably hot, the finger pulsations are large and the vasoconstrictions infrequent. When he is cold the frequency of the vasoconstrictions is greatly increased. The average blood flow of the finger is adjusted to a level appropriate to the necessary heat elimination, not by the maintenance of any steady vascular tone, but by the modification of an underlying rhythmic swinging between constriction and dilatation (Fig. 44).

Bazett and his colleagues (23a, b, c) have studied another important aspect of the circulation in the regulation of temperature. The arteries in the limbs are in effect surrounded by a venous plexus, the venae comites. This anatomical arrangement provides an excellent mechanism for heat exchange (counter-current) between arterial and venous blood. It was shown by Bazett *et al.* that

when the hand is immersed in cold water, the venous blood returning via the venae comites cools the arterial blood, so there may be a
drop of several degrees in the temperature in the brachial and
radial arteries (Fig. 45). They suggested that there may be a
reflex control of the venous circulation, so there is an increased
flow in superficial veins when the limb is warm and an increased
flow in the venae comites in the cold. Such an arrangement would
conserve heat in the cold and increase heat loss in warm conditions.

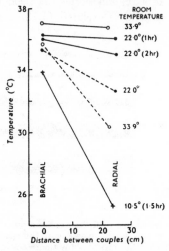

FIG. 45. Intra-arterial temperature measured in the brachial and radial artery:
the distance between the two points is shown on the abscissa. The subject sat at
the room temperature shown for 1–2 hours before measurements were made; the
slope of the lines indicates the gradient of temperature along the artery. The
results shown in dotted lines were obtained when the hand was kept in cold
water. (Redrawn from Bazett, *Am. J. M. Sc.*, **218**, 483, 1949.)

During cold vasodilatation this may not hold good, for it has been
observed that the forearm veins draining blood from the hands are
prominent. The gradient of temperature along these veins from
wrist to elbow is small, suggesting a fairly substantial blood flow
(24). Investigations are needed to measure the proportion of blood
flowing in superficial and deep veins to test Bazett's hypothesis. It
has also been emphasized by Bazett that a consequence of precooling is that there may be a temperature gradient the whole
length of the limbs, instead of only from the skin to the deepest
tissue of the arms or legs. Such a large temperature gradient can
be achieved in the length of a limb, that considerable economy in
heat conservation results.

When the whole body is exposed to a low temperature there is a general cutaneous vasoconstriction and a diminished blood flow in the peripheral tissues. Cooling of one part of the body surface results in a diminished flow in other regions also, and this effect is dependent on an intact vasomotor innervation. Such effects can be demonstrated in the nasal mucosa (25, 5g). There is a rise in rectal temperature initially, during body cooling, as the decreased peripheral blood flow results in a fall in skin temperature, and an increased gradient from the deep tissues to the surface. Heat flow from the core of the body to the superficial layers is therefore reduced, and the temperature rises in the deep tissues. This apparently paradoxical effect will be described in further detail in Chapter 11. Other effects on the circulation of general body cooling, including changes in blood volume and shifts of blood, are discussed in Chapter 10.

Cold Diuresis

The changes in blood flow produced by cold are not confined to the surface vessels, but may involve other organs, including the kidney. It is a matter of common observation that there is an increased output of urine on exposure to cold, and it is sometimes assumed that it is due to a decreased water loss from the skin. However, more detailed studies have shown that this explanation is inadequate, as the diuresis is independent of cutaneous water loss. Bazett *et al.* (26) observed an increased urinary secretion in subjects moved from a warm to a cool environment and found that the increased secretion persisted for several days. During the first few days of exposure to cold there is a gradual fall in plasma volume, and it was suggested that the continued diuresis and decreased blood volume were related. Adolph and Molnar (27) emphasized an interesting feature of cold diuresis, namely the effect of posture. Diuresis is markedly increased on lying down and reduced in the upright position (see Fig. 46). There is an increased chloride content of the urine, and Bazett *et al.* found that their subjects were in negative chloride balance. Stein, Eliot and Bader (28) in a more detailed study confirmed that there was both a negative chloride balance and a negative fluid balance during relatively short exposures to cold.

Recently, Bader, Eliot and Bass (29) have investigated the mechanism of cold diuresis. Their subjects spent five hours reclining, half of the time in a warm room and half in a cool room, and their water intake was maintained at 60 ml./hr. In six out of ten experiments the first $2\frac{1}{2}$ hours were spent lying down covered with blankets in a room maintained at 26°C (80°F) and the last $2\frac{1}{2}$ hours, nude in a room kept at 15·5°C (60°F). In the remaining experiments the order was reversed. There was a consistent increased urinary output in the cold room, averaging 4·76 ml./min. compared with 0·85 ml./min. in the warm room. The total output in the cold room was greater when the warm exposure preceded

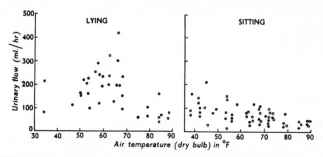

FIG. 46. Rate of urine flow in men in shorts only lying or sitting outdoors, for periods of 2 hours at the temperature shown. There is a considerable diuresis at air temperatures below 21°C (70°F) when the subjects are lying, but it is much smaller in the sitting position. (Redrawn from Adolph and Molnar, *Am. J. Physiol.*, **146**, 507, 1946.)

the period in the cold. The upright posture and also exercise markedly inhibited the cold diuresis. Renal plasma flow was slightly higher in the cold room than in the hot room, but the difference was not statistically significant. Changes in the U/P ratio and in chloride excretion suggested that cold diuresis was due to diminished tubular re-absorption of water. Experiments were carried out to investigate the rôle of posterior pituitary antidiuretic hormone. Cold diuresis was completely inhibited by a dose of 0·5 μg/kg/hr of pitressin, and diminished by 0·1 μg, but 0·001 μg had no effect at all. These doses of posterior pituitary extract are extremely small and demonstrate a very marked sensitivity of the cold diuresis to the anti-diuretic hormone. Bader *et al.* therefore consider that the increased urinary output in the cold is not due to a change in renal blood flow but is regulated by the posterior pituitary.

REFERENCES

1. BARCROFT, H. and EDHOLM, O. G. The Effect of Temperature on Blood Flow and Deep-Temperature in the Human Forearm. *J. Physiol.*, **102**, 5, 1943.
2. LEWIS, T. Observations upon the Reactions of the Vessels of the Human Skin to Cold. *Heart*, **15**, 177, 1930.
3. GRANT, R. T. and BLAND, E. F. Observations on Arteriovenous Anastomosis in Human Skin and in the Bird's Foot with Special Reference to the Reaction to Cold. *Heart*, **15**, 385, 1931.
4. KRAMER, K. and SCHULZE, W. Cold Dilatation in Skin Vessels. *Pflügers Arch.*, **250**, 141, 1948.
5. ASCHOFF, J. (*a*) Research on Temperature Regulation. *Pflügers Arch.*, **247**, 469, 1944.
   (*b*) The Rise of Rectal Temperature during Circumscribed Cooling of Body Surface. *Pfl gers Arch.*, **248**, 149, 1944.
   (*c*) Spontaneous and Reflex Vasomotor Responses in Skin. *Pflügers Arch.*, **248**, 171, 1944.
   (*d*) Cold Vasodilatation. *Pflügers Arch.*, **248**, 178, 1944.
   (*e*) The Cold Dilatation in Ice Water. *Pflügers Arch.*, **248**, 183, 1944.
   (*f*) The Relationship between Temperature and Circulation in the Extremities. *Pflügers Arch.*, **248**, 197, 1944.
   (*g*) Circulatory Regulation of Cold Dilatation. *Pflügers Arch.*, **248**, 436, 1944.
6. CLARA, M. 'Arteriovenous Anastomosis'. J. A. Barth, Leipzig, 1939.
7. SPEALMAN, C. R. Effect of Ambient Temperature and of Hand Temperature on Blood Flow in Hands. *Am. J. Physiol.*, **145**, 218, 1945.
8. (*a*) GREENFIELD, A. D. M. and SHEPHERD, J. T. A Quantitative Study of the Response to Cold of the Circulation through the Fingers of Normal Subjects. *Clin. Sc.*, **9**, 323, 1950.
   (*b*) GREENFIELD, A. D. M., SHEPHERD, J. T. and WHELAN, R. F. The Average Internal Temperature of Fingers Immersed in Cold Water. *Clin. Sc.*, **9**, 349, 1950.
   (*c*) GREENFIELD, A. D. M., SHEPHERD, J. T. and WHELAN, R. F. The Loss of Heat from the Hands and from the Fingers Immersed in Cold Water. *J. Physiol.*, **112**, 459, 1951.
   (*d*) GREENFIELD, A. D. M., SHEPHERD, J. T. and WHELAN, R. F. The Part Played by the Nervous System in the Response to Cold of the Circulation through the Finger Tip. *Clin. Sc.*, **10**, 347, 1951.
   (*e*) GREENFIELD, A. D. M., KERNOHAN, G. A., MARSHALL, R. J., SHEPHERD, J. T. and WHELAN, R. F. Heat Loss from Toes and Forefeet during Immersion in Cold Water. *J. Appl. Physiol.*, **4**, 37, 1951.
   (*f*) GREENFIELD, A. D. M., SHEPHERD, J. T. and WHELAN, R. F. Circulatory Response to Cold in Fingers Infiltrated with Anaesthetic Solution. *J. Appl. Physiol.*, **4**, 785, 1952.
9. WHELAN, R. F. Effect of Cold on the Peripheral Circulation. M.D. Thesis, Queen's University, Belfast, 1951.

10. BLAIR, J. R. 'Cold Injury; Transactions of the First Conference.' P. 51. Josiah Macy, Jr., Foundation, New York (1952).

11. MILLER, H. R. Phasic Fluctuations of Skin Temperature of Fingers and Toes Exposed to Extreme Cold. U.S. Signal Corps. Climatic Research Unit, Fort Monmouth, 1943.

12. BLAISDELL, R. K. Cold Induced Vasodilatation. Report No. 177, Environmental Protection Section, Quartermaster Climatic Research Laboratory, 1951.

13. WOLFF, H. H. and POCHIN, E. E. Vasodilatation-after-reaction in recently Cooled Fingers. *Clin. Sc.*, **8**, 145, 1949.

14. WOLF, S. and HARDY, J. D. Pain Due to Local Cold. *J. Clin. Investigation*, **20**, 521, 1941.

15. BLAISDELL, R. K. Pain and Cold Sensation during Cold Induced Vasodilatation. Report No. 182, Environmental Protection Section, Quartermaster Climatic Research Laboratory, 1951.

16. (a) BADER, M. E. and MEAD, J. Blood Flow in Finger and Wrist at 32°C Room Temperature. *Fed. Proc.*, **8**, 6, 1949.
    (b) MEAD, J. and BADER, M. E. Effect of Body Cooling and Heating on Blood Flow in Finger. Report No. 158, Environmental Protection Section, Quartermaster Climatic Research Laboratory, 1949.

17. RAPAPORT, S. I., FETCHER, E. A., SHAUB, H. G. and HALL, J. F. Control of Blood Flow to the Extremities at Low Ambient Temperature. *J. Appl. Physiol.*, **2**, 61, 1949.

18. BURTON, A. C. (a) Proper Distribution of Heat to Different Parts of the Body in Electrically Heated Flying Suits. Report No. C 2138, Associate Committee on Aviation Medical Research, 1942.
    (b) Tests of the Thermal and Electro-Thermal Properties of the Taylor Suit. Report No. C 2045, Associate Committee on Aviation Medical Research, 1942.

19. BENTLEY, A. N., BURTON, A. C., KITCHING, J. A., NOAKES, F. and PAGÉ, E. Heat Regulator for the Automatic Regulation of Electrically Heated Clothing. Report No. C 2401, Associate Committee on Aviation Medical Research, 1943.

20. BARCROFT, H. and EDHOLM, O. G. Temperature and Blood Flow in the Human Forearm. *J. Physiol.*, **104**, 360, 1946.

21. BADER, M. E. and MACHT, M. B. Indirect Peripheral Vasodilatation. *J. Appl. Physiol.*, **1**, 215, 1948.

22. BURTON, A. C. and TAYLOR, R. M. A Study of the Adjustment of Peripheral Vascular Circulation to the Requirements of the Regulation of Body Temperature. *Am. J. Physiol.*, **129**, 565, 1940.

23. (a) BAZETT, H. C., LOVE, L., NEWTON, M., EISENBERG, L., DAY, R. and FORSTER, R. Temperature Changes in Blood Flowing in Arteries and Veins in Man. *J. Appl. Physiol.*, **1**, 3, 1948.
    (b) BAZETT, H. C., MENDELSON, E. S., LOVE, L. and LIBET, B. Precooling of Blood in the Arteries, Effective Heat Capacity and Evaporative Cooling as Factors Modifying Cooling of the Extremities. *J. Appl. Physiol.*, **1**, 169, 1948.
    (c) LOVE, L. Heat Loss and Blood Flow of the Feet under Hot and Cold Conditions. *J. Appl. Physiol.*, **1**, 20, 1948.

24. COOPER, K. E. and EDHOLM, O. G. Unpublished Observations, 1950.

25. MUDD, S., GOLDMAN, A. and GRANT, S. B. Chilling of the Body Surface, Reflex Vasoconstriction in Mucous Membrane of Palate, Pharynx, etc. *J. Exper. Med.*, **34,** 11, 1921.

26. BAZETT, H. C., SUNDERMAN, F. W., DOUPE, J. and SCOTT, J. Climatic Effects on the Volume and Composition of Blood in Man. *Am. J. Physiol.*, **129,** 69, 1940.

27. ADOLPH, E. F. and MOLNAR, G. W. Exchanges of Heat and Tolerances to Cold in Men Exposed to Outdoor Weather. *Am. J. Physiol.*, **146,** 507, 1946.

28. STEIN, H. J., ELIOT, J. W. and BADER, R. A. Physiological Reactions to Cold and their Effects on the Retention of Acclimatization to Heat. *J. Appl. Physiol.*, **1,** 575, 1949.

29. BADER, R. A., ELIOT, J. W. and BASS, D. E. Hormonal and Renal Mechanisms of Cold Diuresis. *J. Appl. Physiol.*, **4,** 649, 1952.

CHAPTER 9

THE METABOLIC RESPONSE TO COLD

The knowledge that the heat production of animals increases in the cold must be very old. Lavoisier (1) in his experiment with Laplace in 1780 on the guinea-pig in the ice-calorimeter, which was the birth of animal calorimetry, recognized that there was an error in his proof that the heat given off by the animal was due to the oxidation of food. This error was that the experiment in which they measured the heat from the animal, by the amount of ice melted, was at a different temperature from that in which they measured the CO_2 output. Lavoisier recognized that the respiration in the cold might be greater. The greatest advances in study of the increase of metabolism to cold were made by Rubner (2), who classified the temperature regulatory responses of the animal under the two headings of 'physical regulation', control of heat loss by peripheral circulation and by sweating, and 'chemical regulation', control of heat production. For a full account of the earlier work, the relevant chapters in the book by Lusk (3) should be read.

Under the term 'chemical regulation' Rubner included an increase of the metabolism of body tissues of all varieties in response to cold. There has never been any disagreement as to the existence of this increase, but for more than 50 years there has been debate whether the increase is mainly or even entirely from increased muscular activity, or whether other tissues than muscle increased their metabolic rate also in response to cold. The tide of opinion has ebbed and flowed. Johansson (4) and Sjöstrom (5), contrary to the view of Voit (6) and of Rubner, were convinced that in man the whole of the additional metabolism was due to shivering. Loewy (7), Morgulis (8) and Benedict (9) agreed with these authors that increases in metabolism took place only when accompanied by shivering.

On the other hand, those who emphasized the importance of 'extra-muscular' increases of metabolism stated that they measured increases of metabolism in man and in animals up to 40 per cent, 'without any accompanying shivering'. The work of Cannon *et al.* (10) on the secretion of adrenaline in animals that were cooled, and his belief in the general calorigenic effect of adrenaline, led him to state that extra-muscular chemical regulation was explained by the action of this hormone. However, there now seems to be considerable doubt as to the calorigenic effects of adrenaline (11).

The crux of the matter is that the supposed proof of the participation of tissues other than muscle rests upon the evidence of increases of metabolism to cold *'without shivering'*. When it is realized that there might be increases of muscle activity in the cold, i.e. increased muscle tone, not detectable as gross shivering, which implies a tremor of the muscle, it is evident that a careful study of the nature of the activity in muscles, provoked by cold, is necessary. Swift (12) studied the increased metabolism in man, and was convinced that any increase in metabolism not accompanied by definite shivering might be justly ascribed to increased muscular tension. One of his subjects, at room temperature, imitated as faithfully as possible by voluntary effort the increase in tension that he had experienced in experiments in the cold, and there was a 36 per cent increase in oxygen consumption. On the other hand, DuBois and Hardy (13) found an earlier rise in metabolism to cold in women than in men, without frank shivering, and consider muscle tension inadequate to account for this.

Mechanism of Shivering

Further evidence of the existence of a 'thermal muscular tone' which preceded the full muscular activity that would be described as shivering, was provided by the use of electromyographic recording from the muscles of cooled cats (14). Needle electrodes of the well-known type were thrust into various muscles of the hind limbs of lightly anaesthetized cats, so that the action currents of single muscle units could be recorded, and the over-all activity of the muscles estimated by the noise heard in a loud speaker, while the activity of single units could be judged by the frequency of their discharge. The animals were cooled at a controllable rate by circulating cold water through a metal 'bullet' on the end of a stomach catheter. The typical response is shown in Fig. 47. In the anaesthetized cat the muscles are normally completely relaxed and no action currents are recorded or heard. Unless the room is very warm the rectal temperature falls steadily. At a critical level of deep body temperature, determined very largely by the depth of anaesthesia, single muscle units are heard and recorded in intermittent activity, firing at frequencies of as little as 5 per second.

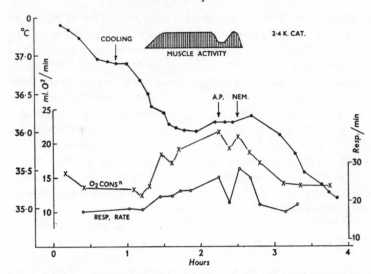

FIG. 47. Typical response in O_2 consumption, respiratory rate, and muscle activity, from needle electrode in gastrocnemius muscle of a cat anaesthetized with nembutal. Cooling was by stomach tube. A.P. Injection of 1 ml. 10% antipyrine intravenously. Nem. 0·5 ml. of 5% Nembutal intravenously. (From unpublished work of Burton and Bronk.)

The activity of such units rapidly increases in frequency of discharge, and the number of units that are active also increases. As the figure shows, the appearance of this electrical activity of the muscles coincides with the first increase in oxygen consumption and in respiratory volume. The figure also shows how the muscular activity can be suppressed by the injection of an antipyretic drug or of more anaesthetic. The changes in rectal temperature show the effectiveness of the increased muscle activity in producing heat which slows or prevents a further fall of body temperature.

Examination of the records of the action currents shows that in the initial stages the muscle units fire at different frequencies of repetition and out of phase with each other; this represents an increased muscle tone without any co-ordinated tremor. Only in later stages do the discharges approach synchrony, so that there are groups of discharges repeated at the frequency of the tremor, which in the muscles of the cat is from 10 to 12 per second (Fig. 48). The mechanism of this eventual co-ordination of the various motor units was found to be the periodic inhibition of the impulses in the motor nerve, rather than by the central synchronization of these impulses. For example, Fig. 48 shows how the profile of successive groups of impulses, as they escape this periodic inhibition, changes from group to group, indicating that even at this stage the individual motor units are firing at different frequencies.

Attempts to record action currents in the cord with a rhythm the same as that of the muscle tremor were entirely negative, and there seems no doubt that the genesis of the tremor is by the inhibitory effects of periodic discharges in the afferent fibres from

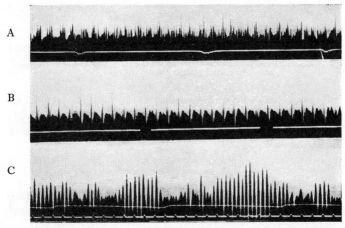

A

B

C

Fig. 48. Stages in the development of co-ordinated activity and tremor in muscle units. The earliest stage when individual units are completely out of phase is not shown. A. Some grouping of impulses is seen but units are at very different frequencies. Time intervals 1 second. B. Complete grouping of impulses by the periodic inhibition of all impulses. Time intervals 1 second. C. Violent shivering showing the characteristic waxing and waning. The record is derived from artefact due to movement of the electrode needle. Time intervals 1/5 second. (From unpublished work of Burton and Bronk.)

the muscle, i.e. due to an oscillatory character of the motor nerve-afferent arc system. Cutting the dorsal roots removed the tremor of shivering but not the thermal muscular tone. Perkins (15) has shown that the frequency of tremor in neighbouring muscles is not the same, and can be altered by mechanical loading of the limb. Similar conclusions as to the nature of the tremor were reached by Denny Brown *et al.* (16) who recorded action currents and the mechanical tremor of various muscle groups in man.

Anyone who has observed closely the shivering in animals or man must have noted that characteristically it waxes and wanes in intensity (Fig. 48). A major part of the fluctuation is obviously

L

related to the respiration. An explanation for the respiratory rhythm is provided in the spinal reflex effects that are shown by a muscle in thermal muscular tone. For example, if the tendon of a muscle, from which the activity of shivering is being recorded by a needle electrode, be stretched by loading with a small weight, the activity is greatly facilitated (Fig. 49). Simultaneously the activity in the antagonist muscle will be partially or totally inhibited. All of the spinal reflex effects, both ipsilateral and

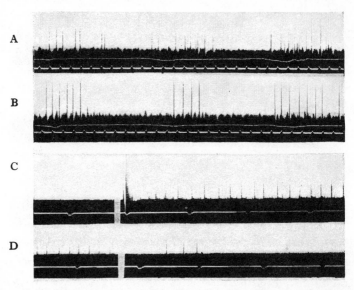

FIG. 49. Action currents recorded by a needle electrode in the muscles of a cooled cat. A and B, respiratory fluctuation in the quadriceps. The white line below is the respiration, inspiration being up. Time intervals are 1/5 second. C and D, facilitation of shivering in the gastrocnemius muscle, when stretched by 5 gm. on the tendon at the vertical white line. The brief stretch reflex is followed by a steady train of impulses as long as the weight is left on (until the second vertical white line). (From Burton and Bronk—unpublished.)

contralateral, that were shown by Sherrington and his pupils on a background of activity due to decerebrate rigidity, can easily be demonstrated on the muscles of a cooled animal, where thermal muscular tone provides the tonic activity. It is therefore easy to explain a marked respiratory rhythm, for example, in the thigh muscles (Fig. 49), as due to change in the stretch of those muscles when the diaphragm moves in the respiratory cycle.

In addition to the respiratory rhythm in shivering, there is a

periodic fluctuation of longer duration, which experimental subjects lying shivering in a cold room have described as 'bouts of
shuddering or shaking'. This may well be related to the changes
in skin temperature, and therefore in the intensity of afferent
stimulation to the regulatory centre, that occur over the muscles in
shivering. Before a bout of shivering the skin temperature falls
slowly but rises during and after each bout, and a distinct alterna-
tion of intensity of sensation of cold can be noticed.

Shivering, as part of the temperature regulatory mechanism, is
an autonomic or involuntary function of the body. It should therefore not astonish us to learn that, while the final neurone involved
is of course the same motor nerve, the pathway in the cord is not
the pyramidal tract responsible for voluntary muscular activity.
This was shown by Uprus *et al.* (17) by examining patients with a
variety of lesions of the spinal cord. It was possible to find some
who had paralysis of voluntary muscle movement, yet shivering
could be elicited in these muscles. They concluded that the pathway was by the lateral tecto-spinal or rubro-spinal tracts. Shivering
in a given muscle tends to be suppressed during voluntary activity
of that muscle.

Shivering may be considered, from the thermodynamic point of
view, to be more efficient than voluntary contraction for the production of heat, the function it evidently serves. While 60 to 70 per
cent of the energy consumed in a voluntary contraction appears
immediately as heat (i.e. the thermodynamic efficiency of muscle
is 30 to 40 per cent), the rest is not immediately available as heat
to keep the animal warm, unless by internal friction in the muscle
or friction in the environment the mechanical work is transformed
into heat. In shivering there is immediate transformation of all
the energy involved into heat, as the limbs do not grossly move.
Because agonist and antagonist both increase in tension, no external
work is done. The mechanism in the muscle probably represents
one which produces the least tension for the most energy and heat
production, and interferes least with voluntary muscular activity.

Amount of Increase in Metabolism in the Cold

Many examples of the progressive increase in heat production of
animals, when the environmental temperature is below the 'critical
temperature' for that species, have already been discussed in
Chapter 6 in connexion with the work of Irving *et al.* (see Figs.
24, 25 and 26). Comparable data for man are not available, for

we would require that a series of experiments be made at different environmental temperatures on men in the same clothing at all temperatures. Data on naked or very lightly clad men are to be found in the calorimetric studies of Gagge *et al.* (18) and of Hardy and DuBois (13). Since these workers use different temperature scales (Gagge *et al.* use 'operative temperatures') and conditions of air movement, etc., are also not standard, it is not easy to decide what is the 'critical temperature' for man at which

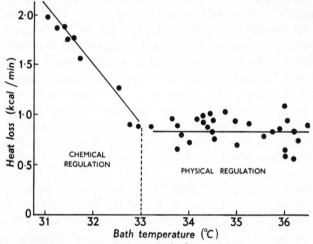

FIG. 50. Heat loss in a well-stirred water bath at different temperatures. The periods taken were in the steady state, so the heat production must have been the same as heat loss. (Redrawn from Burton and Bazett, *Am. J. Physiol.*, **117,** 1, 1936.)

heat production begins to increase, but it is in the neighbourhood of 28°C (82°F) for the naked man. This would substantiate the suggestion of Irving and his co-workers that man is a 'tropical' animal. For man immersed in water baths at different constant temperatures Burton and Bazett (19) found a critical water temperature of 33°C (91°F) as shown in Fig. 50.

Factors affecting Shivering and Thermal Muscular Tone

Like the rest of the temperature regulatory reflexes, shivering is extremely susceptible to depression by anoxia. This can be shown electromyographically if the animal breathes nitrogen rather than air. The decrease in frequency and number of single motor units is very abrupt, (Fig. 51) and recovery is prompt when air is restored. It is interesting that air-crews almost universally report that the breathing of oxygen makes them feel warmer, even at low altitudes (10,000 ft.). Antipyretic drugs also suppress shivering, and there seems to be a special inhibition of shivering by magnesium ions (20). This cannot be explained as due to the well-known curare-like effects of this ion when in sufficient concentration in the blood, for it occurs at much lower concentrations than this. Vasodilatation occurs, but at the same time panting is stimulated, presumably by an effect on the centre, so we cannot

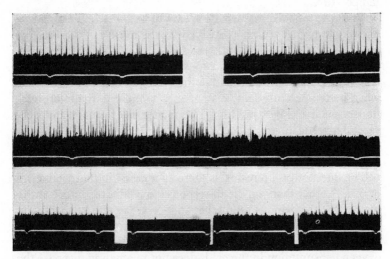

Fig. 51. Cessation of shivering impulses, recorded by a needle electrode in the gastrocnemius muscle of a cooled cat, when nitrogen was substituted for air to breathe (at the first interruption of the record). Time intervals are seconds, and the second and third lines of record are from a continuous record. The intervals at the interruptions in the third line are each of 1 minute, after O_2 had been restored. (From Burton and Bronk, unpublished.)

regard the antipyretic action of magnesium merely as a case of general C.N.S. depression. It is of great interest that this ion seems to have a special relation to hibernation (21, 22, 23).

Shivering is promptly inhibited by the injection of insulin (24), presumably through the lowering of blood sugar. Injection of insulin also will cause a hibernant to go into the hibernating state, when the conditions are right (25). An inhibition of shivering when breathing 1–5 per cent CO_2 has been reported by Miller *et al.* (26).

The sensitivity of the temperature regulatory mechanisms to depression by general anaesthesia is well known. Hemingway (27) showed that the critical body temperature at which shivering appeared in cooled dogs was related to the dosage of barbital and shivering was more sensitive to depression than the vascular response to cold. The same author has made an excellent study of the standard responses, including shivering of unanaesthetized dogs to cold (28), showing how skin temperature and rectal temperature both control their appearance. Indeed, anaesthesia has often been used to enable experimental hypothermia to be produced, as otherwise violent shivering may prevent the fall of body temperature.

The R.Q. of shivering has been said to be unity, indicating that carbohydrate fuel is used exclusively by the muscle in this activity, and much of the older work has indicated a rise of R.Q. when shivering occurs in cooling experiments. However, changes in respiratory minute volume and in the solubility of CO_2 in the blood when its temperature falls, may greatly complicate the interpretation of such changes in R.Q. The matter needs more and very careful investigation.

Extra-Muscular Chemical Regulation

The question still remains, is there an increase in the metabolism of tissues other than muscle in response to cold? It is important to realize that the demonstration that an increase in muscular activity, which may be called 'thermal muscular tone' and which precedes frank shivering and occurs coincidently with the first rise of oxygen consumption in the cold, does not in any way disprove the existence of an extra-muscular response. It merely invalidates the argument of those who claimed that there could be increases of metabolism up to 40 per cent 'without shivering'. It must be admitted that the question is still unanswered. There is accumulating suggestive evidence along several different lines that after continued exposure to cold in laboratory animals, the liver may have an increased metabolism that could contribute greatly to the total heat production, and possibly replace the response in muscular activity. This evidence is discussed in the chapter on acclimatization. We still lack any evidence whether or not the acute response, in an unacclimatized animal, is from other tissues than muscle. Probably only studies of the heat production of the viscera directly will supply the answer, for even if the alternative approach of eliminating the muscle response by curare-like drugs were used, the results would not be unequivocal. If an increase of total metabolism to cold still persisted, it would be difficult to prove that the muscle response was completely inhibited. If no response to cold was seen after curare, it might be argued that the drug had also inhibited the response of other tissues than muscles.

Nutrition in the Cold

There is, of course, another way of estimating the effect of environmental cold on heat production in man and animals, by studying their caloric intake in food, and there has been a good deal of evidence of this kind. The calorie value of the diet must be studied over a considerable period of at least several days, to justify interpretation in terms of heat production, since in the first few days of exposure to cold there is a marked imbalance between intake and output of energy (29). The appetite increases immediately while the heat production increases most slowly, and a weight gain is usually seen in man. In animals, however, according to the severity of the cold conditions, there may be a loss of weight, so that it cannot be assumed that the caloric intake was not supplemented in any given period by burning of tissues, or on the other hand storage in the body.

Much of the data on caloric intake on animals in the cold are discussed in the later chapter on acclimatization. A striking relation between the calories consumed by soldiers, engaged on similar tasks, and the environmental temperature has been shown by Johnson and Kark (30) in Fig. 52. Dietary requirements of man in a cold environment were studied in detail during the war. The authors considered that this great difference could not be explained by differences in basal metabolic rate. They accepted the figure of 20 per cent difference as being the greatest to be observed between basal metabolic rates in the tropics and the Arctic, which would only account for some 460 kcal per 24 hours. It would be easy to place an interpretation on the remarkable relation shown in Fig. 52 that was much too simple. The increase in the cold cannot be due to the increased heat loss from the body, because the points on the graph represent figures for men where the clothing insulation was very different in the separate climates, and this might have been expected to destroy the correlation between total calories and the environmental temperature. Johnson and Kark suggest that the metabolic cost of similar activities was greater in the cold than in temperate or tropical climates, and it is certainly true that every task seems to take more time and effort in the cold. They also suggest that the hobbling effect of the heavy clothing worn in the

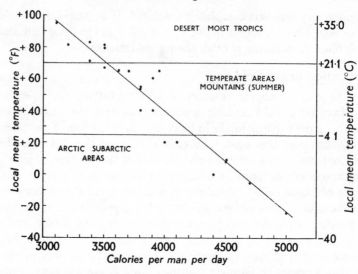

Fig. 52. Voluntary calorie intake of soldiers stationed in climates of different temperatures. (Johnson and Kark, *Science*, **105**, 378, 1947.)

cold environments was in part responsible for the increased energy expenditure. However, Gray, Consolazio and Kark (31) investigated this, with subjects working on a bicycle ergometer at three different temperatures and wearing light, intermediate or heavy clothing, at each of the three temperatures. They found that the calorie requirement (from O_2 consumption) for a given amount of external work increased about 5 per cent when the clothing was changed from very light to 'temperate' clothing, and 5 per cent more again when clothing was changed from temperate to Arctic clothing. Comparing experiments at three different temperatures $-8°C$ (15°F), 15·5°C (60°F) and 32°C (90°F) where the same clothing was worn, the energy cost of the work decreased by only 2 per cent for each change in temperature. This result for the increased cost of work in heavy clothing agrees with the figure of 7 per cent for heavy flying clothing in similar experiments made during the war by the Toronto group. However, it is difficult to see how the increase due to clothing is sufficient to explain more than a small part of the increase from 3,000 to 5,000 kcal per 24 hours shown by the nutritional data. If we assumed a basal value of 1,700 kcal only, and therefore 1,300 kcal for the work in the warm environments, the latter would increase only to about 1,430 kcal in the cold due to the hobbling effect of clothing, and the new total of 3,130 kcal would be still far short of what was found. Either we must conclude that the hobbling effect of clothing for work on a bicycle ergometer falls far short of representing that in doing the usual soldier's daily 'chores', which may well be, or we must look for some other explanation. Careful field studies of the time spent on specific tasks and measurements of the oxygen consumption might elucidate the matter. There are also many energy consuming tasks, like putting on and off heavy boots, gloves and parkas in the cold, that are not performed at all in the warmer environments. Possibly, on the other hand, we are to regard the correlation between food intake and environmental temperature as primarily due to a stimulating effect of cold on the appetite.

Against any superficial interpretation of the relation shown in Fig. 52, it may also be pointed out that according to the curves of metabolism vs. environmental temperature obtained by Irving *et al.* (Chapter 6), and the short-term experiments on man, and according to the theory explaining these, the metabolism should not rise until a critical temperature was reached. In the nutritional data the daily metabolism apparently increases steadily from tropical to Arctic environments.

Summary

The increase of heat production by metabolism as a defence against cold, called by Rubner 'chemical regulation', has been extensively studied in man and in animals, by oxygen consumption measurements and by observations of the nutritional calorie requirements. Attention has of late years been focused on the activity of the muscles in shivering for providing this extra metabolism. Study of the detailed motor mechanism of shivering shows that it is preceded by an increased 'thermal muscular tone', which would not be detected as frank shivering.

While this destroys the argument of some that increased metabolism of other tissues than muscles must be involved because they observed increases of total metabolism without shivering, it leaves the question of extra-muscular metabolic increments in the cold completely open, at least for short term experiments when acclimatization is not concerned.

The voluntary dietary intake of calories by soldiers shows a remarkable increase as the environmental temperature decreases. No simple explanation of this is acceptable, since clothing varied greatly in the different climates in which the data were collected.

REFERENCES

1. LAVOISIER, A. L. and LAPLACE, P. S. *Mem. Acad. Sc.*, 379, 1780.
2. RUBNER, M. 'The Laws of Energy'. Deuticke, Leipzig, 105, 1902.
3. LUSK, GRAHAM. 'The Elements of the Science of Nutrition.' 4th ed. Particularly Chapters 1 and 6. Saunders, Philadelphia and London, 1928.
4. JOHANSSON, J. E. The Influence of Environmental Temperature on the CO_2 Production of the Human Body. *Scandinav. Arch. Physiol.*, 7, 123, 1897.
5. SJOSTROM, L. The Effect of Ambient Air Temperature on the CO_2 Production of Man; A Contribution to the Theory of Temperature Regulation. *Scandinav. Arch. Physiol.*, 30, 1, 1913.
6. VOIT, C. The Effect of Air Temperature on Catabolism in Warm Blooded Animals. *Ztschr. f. Biol.*, 14, 57, 1878.
7. LOEWY, A. The Effect of Cooling on Metabolism in Man. *Pflügers Arch.*, 46, 189, 1890.
8. MORGULIS, S. The Effect of Environmental Temperature on Metabolism. *Am. J. Physiol.*, 71, 49, 1924.
9. BENEDICT, F. G. Factors affecting Basal Metabolism. *J. Biol. Chem.*, 20, 263, 1915.
10. CANNON, W. B., QUERIDO, A., BRITTON, S. W. and BRIGHT, E. M. The Rôle of Adrenal Secretion in the Chemical Control of Body Temperatures. *Am. J. Physiol,.* 79, 466, 1926.
11. GRIFFITH, F. R. Fact and Theory Regarding the Calorigenic Action of Adrenaline. *Physiol. Rev.*, 31, 151, 1951.
12. SWIFT, R. W. The Influence of Shivering, Subcutaneous Fat, and Skin Temperatures on Heat Production. *J. Nutrition*, 5, 227, 1932.
13. HARDY, J. D. and DUBOIS, E. F. Differences between Men and Women in their Response to Heat and Cold. *Proc. Nat. Acad. of Sc.*, 26, 389, 1940.
14. BURTON, A. C. and BRONK, D. W. The Motor Mechanism of Shivering and Thermal Muscular Tone. *Am. J. Physiol.*, 119, 284, 1937.
15. PERKINS, J. F. The Rôle of the Proprioceptors in Shivering. *Am. J. Physiol.*, 145, 264, 1946.
16. DENNY-BROWN, D., GAYLOR, J. B. and UPRUS, V. Note on the Nature of the Motor Discharge in Shivering. *Brain*, 58, 233, 1935.
17. UPRUS, V., GAYLOR, G. B. and CARMICHAEL, E. A. Clinical Study with Especial References to Afferent and Efferent Pathways. *Brain*, 58, 220, 1935.
18. GAGGE, A. P., WINSLOW, C.-E. A. and HERRINGTON, L. P. Influence of Clothing on Physiological Reactions of the Human Body to Varying Environmental Temperatures. *Am. J. Physiol.*, 124, 30, 1938.
19. BURTON, A. C. and BAZETT, H. C. A Study of the Average Temperature of the Tissues, of the Exchanges of Heat and Vasomotor Response in Man by Means of a Bath Calorimeter. *Am. J. Physiol.*, 117, 36, 1936.
20. HEAGY, F. C. and BURTON, A. C. Effect of Intravenous Injection of Magnesium Chloride on the Body Temperature of the Unanaesthetized Dog, with some Observation on Magnesium Levels and Body Temperature in Man. *Am. J. Physiol.*, 152, 407, 1948.
21. SUOMALAINEN, P. Magnesium and Calcium Content of Hedgehog Serum during Hibernation. *Nature*, 144, 443, 1939.
22. SUOMALAINEN, P. Production of Artificial Hibernation. *Nature*, 142, 1157, 1938.

23. SUOMALAINEN, P. Artificial Hibernation. *Nature*, **144,** 443, 1939.

24. FINNEY, W. H., DWORKIN, S. and CASSIDY, G. J. The Effects of Lowered Body Temperature and of Insulin on the Respiratory Quotient of Dogs. *Am. J. Physiol.*, **80,** 301, 1927.

25. DWORKIN, S. and FINNEY, W. H. Artificial Hibernation in the Woodchuck (Arctomys Monax). *Am J. Physiol.*, **80,** 75, 1927.

26. MILLER, H. R., GRUNDFEST, H., ALPER, J. M., MARGOLIN, S. G., KORR, I. M., FEITELBERG, S. and KLEIN, D. Inhibition of Shivering by Inhalation of 1–5 per cent CO_2, with Increased Feeling of Warmth. Climatic Research Unit, Fort Monmouth. Report 13 CR. 1944.

27. HEMINGWAY, A. The Effect of Barbital Anaesthesia on Temperature Regulation. *Am. J. Physiol.*, **134,** 350, 1941.

28. HEMINGWAY, A. The Standardization of Temperature Regulatory Responses of Dogs to Cold. *Am. J. Physiol.*, **123,** 736, 1939.

29. BURTON, A. C., SCOTT, J. C., McGLONE, B. and BAZETT, H. C. Slow Adaptations in the Heat Exchanges of Man to Changed Climatic Conditions. *Am. J. Physiol.*, **129,** 84, 1940.

30. JOHNSON, R. E. and KARK, R. M. Environment and Food Intake in Men. *Science.*, **105,** 378, 1947.

31. GRAY, LeB., CONSOLAZIO, F. C. and KARK, R. M. Nutritional Requirements for Men at Work in Cold, Temperate and Hot Environments. *J. Appl. Physiol.*, **4,** 270, 1951.

CHAPTER 10

ACCLIMATIZATION TO COLD

In this chapter an account will be given of the changes that occur in animals and man when they are exposed to low temperatures for long periods of time. The acute adjustments are described in detail in Chapter 11, but slower and more long-lasting effects can be discerned when animals live in a cold environment.

There are a number of terms in use to describe these chronic effects, and definitions have not been generally agreed. Hart (1) has suggested these definitions.

Acclimatization   Changes in the responses of the organism produced by continued alterations in the environment.

Acclimation   Alterations related to changes in a life-time.

Adaptation   Changes occurring during a period of several generations.

The ugly word 'accustomization' has been proposed to describe the changes in the mode of living that have to be made when man moves to an Arctic environment. It seems an unnecessary addition to the list of repulsive jargon, when there are many reasonable alternative phrases. 'Technique of Arctic living', 'habituation', 'experienced', 'self-adaptation', 'naturalization', etc.

It is certainly important to have a word or phrase which emphasizes the part played by the technique of living when men become acclimatized to Arctic conditions. This Arctic training is probably the major adjustment which occurs in the individual. The term 'habituation' is therefore proposed as an alternative to 'accustomization'.

The definitions above include 'adaptation'. As this word has been used in biology in a strict sense for a long time, it should not be used to describe short term changes. It must be admitted that physiologists use the term to describe relatively rapid changes, i.e. dark adaptation, adaptation of receptors. In this book 'adaptations' will only be used in the strict biological sense.

This chapter will be mainly concerned with acclimatization, although adaptation in animals will also be described. Much of the work to be described will deal with animal experiments, as the results obtained so far in man are in general equivocal.

Acclimatization in Animals

Interesting observations have been made on fish by Fry (2). Fish acclimatize rapidly to a wide range of temperatures, so the

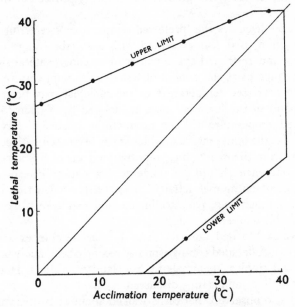

FIG. 53. The thermal tolerance of goldfish. The fish is acclimated to a particular temperature (Acclimation Temperature) and the upper and lower lethal limits determined. A fish acclimates to 10°C, has an upper limit of 30°C and a lower limit of 0°C, but if it is acclimated to 25°C, it can tolerate temperatures up to 36°C, but will not survive at temperatures of 5°C or lower. (From Fry, *Pub. Ont. Fish Research Lab.*, 1947.)

upper and lower lethal levels can be plotted as different points according to the thermal history of the fish (Fig. 53). The figure shows the thermal tolerance of the goldfish: as the upper level of lethal temperature rises, so does the lower level, and the tolerance limit persists at approximately 30°C (54°F) range. The area enclosed by the upper and lower lethal limits represents the temperature range within which the acclimatized fish can live indefinitely. Beyond this area of tolerance there is a zone of resistance, temperatures which will ultimately be lethal but in which the animal can survive for a time. The rate of acclimatization varies considerably in different species of fish.

Whether similar types of relationships hold good in warm blooded animals has yet to be demonstrated. However, it is valuable in considering warm blooded animals to be aware of these relationships, particularly the problem of the gain in heat tolerance which is accompanied by loss of cold tolerance and vice versa. It is not clear if these two processes necessarily proceed at the same rate.

Fry also discusses the 'preferred temperature', referring to the phenomenon 'that fish presented with a suitable range of temperature and restricted space will tend to congregate at one end of that range or at some more or less definite temperature within it'. The preferred temperature of the biologist may be taken to correspond to the 'comfort zone' as defined by Yaglou (3). The preferred temperature will depend upon the thermal history of the animal, i.e. the temperature to which it has been acclimatized.

Apart from the many other important aspects of Fry's work, for which reference should be made to his paper 'Effects of the environment on animal activity', it is clearly shown that in cold blooded animals, e.g. fish, acclimatization can take place over a considerable temperature range.

In warm blooded animals there is considerable evidence of adaptation. A detailed comparison has recently been made between Arctic and tropical mammals by Scholander, Walters, Hock and Irving (4a, b, c), and is fully discussed in Chapter 6.

They also observed that birds walked about at temperatures of $-40°$ to $-50°C$ without frostbite. But a gull which was kept indoors at $+ 20°C$ escaped into the snow at $-20°C$ and the web of the feet froze hard within a minute with subsequent gangrene. This might be considered as an example of loss of acclimatization.

Irving (5) has summarized these important studies as discussed in Chapter 6. Climatic adaptation can be achieved by variation in body temperature, insulation and metabolism. There is no evidence that deep body temperature varies significantly in mammals in different parts of the world. Body temperatures were measured in Arctic animals in the winter immediately after they were shot in the bush and averaged $37°C$ ($98·6°F$). Reindeer and dogs living at temperatures of $-45°C$ ($-49°F$) had normal rectal temperatures, so it can be concluded that adaptation to cold is not achieved by a lowering of body temperature. The metabolic cost of living in the Arctic is stated to be about the same as in warmer climates. This statement is based on the findings that the basal metabolic rate in the Arctic animals is proportional to size and weight. 'The points fall on the mouse–elephant curve'. This must not be interpreted to mean that the total energy expenditure per diem is the same in the cold as in the heat.

Irving concludes that insulation is the most important factor of adaptation to cold in Arctic animals.

Changes in Animals Exposed to Cold

A large volume of work has been done on the effects of exposing laboratory animals to different ambient temperatures. The rat, guinea-pig and rabbit have been studied in the greatest detail.

Metabolic rate increases when animals are exposed to cold, depending on the critical temperature of the animals (*vide supra*). This increase in metabolism is, in part at least, due to increased muscular activity in the form of shivering. Shivering is preceded by an increased muscle tone, accompanied by increased action potentials. The increase in metabolic rate due to shivering may be as high as 6–7 times the basal metabolic rate, for short periods (see Chapter 8).

However, in animals maintained at low temperatures for long periods, metabolic rate is kept at a high level without gross shivering being apparent. Blair *et al.* (6) studied rabbits at −30°C (−22°F) for many weeks. Food intake increased from 60 g per day at 23°C (73·4°F) to 105 g per day, with no appreciable change in weight. The rabbits showed no apparent change in activity in the cold. They hunched up in a ball with their ears tucked into their fur. The fur was erected at right angles to skin surface and had a covering of delicate hoar-frost. No gross shivering could be observed although there was a very fine tremor. Rectal temperature did not change significantly. Such observations do not provide complete evidence of a chemical regulation as opposed to shivering, but do suggest that the increased muscular activity may not be adequate to explain the increased food intake. Chinn *et al.* (7) found that food intake increased 60 per cent in rats kept at 4°C (39°F) for 10 days. Shivering appeared to be minimal or absent.

Dugal, whose studies will be described in detail below, has carefully followed the weight changes in rats exposed to cold. He uses young, growing animals: on exposure to cold the weight drops and this period of initial loss of weight he terms the state of resistance. This is followed by a slow recovery of weight to the initial level, which is described as the state of acclimatization, then growth continues at the pre-cold rate and he calls this the state of adaptation.

Summarizing our knowledge so far on the effect of exposure to cold; animals show an increased metabolic rate when exposed below a critical temperature. This critical temperature largely depends on the control of insulation and the degree of insulation possessed by the animal and therefore the critical temperature varies for different species. The increase in metabolism is accompanied by an increased food intake, so that in many animals there is no change in weight. However, the metabolic increase may be greater than the food intake and therefore the weight may decrease. The change in metabolism has a very important muscle component. There is suggestive evidence that in addition to increased muscle metabolism, other tissues also increase their metabolism owing to the activity of the thyroid, suprarenal, etc.

It has been shown by many workers that animals exposed to the cold for long periods have marked changes in the thyroids and suprarenals. Histologically, the thyroid shows increased activity in animals exposed to cold (8). LeBlond *et al.* (9) gave radio-active iodine to three groups of rats kept at 0° to 2°C (32° to 35·6°F), 19° to 25°C (66° to 77°F) and 32° to 34°C (90° to 93°F). They were sacrificed at varying periods. After 7 days there was marked histological evidence of thyroid stimulation in 5 out of 7 animals kept in the cold. Four out of 5 animals showed similar evidence of thyroid stimulation after 26 days in the cold. The thyroid fixed and metabolized iodine greatly in excess of the controls during the period of 7 to 26 days. After 40 days, iodine metabolism was back to control levels and the histological appearance of the thyroids was similar to the controls. On the other hand, hypophysectomized animals do not show any histological evidence of increased thyroid activity in the cold.

Prolonged exposure to cold also causes an increased weight of the suprarenals. Although most authors find that the increase takes place in the cortex, Morin (10) and Schaeffer (11) consider that the medulla is also enlarged. Thérien (12) concludes that the hyperactivity of the thyroid, suprarenal and probably the hypophysis indicates that there is an increased metabolism in the cold which can be attributed to chemical regulation. Mansfeld has shown that there is a seasonal difference in response to thyroxine in animals. The effect of thyroxine diminishes in the warm season and increases in the cold, and this effect is said to be due to the secretion in excessive heat of thermothyrine which inhibits the metabolic effect of thyroxine (13, 14).

When rats are exposed to low temperatures there are a number of responses which are similar to those aroused by other forms of stress. These changes are termed by Selye (15) the alarm reaction. Continued exposure to cold leads to the stage of adaptation and may terminate in the stage of exhaustion. The numerous publications by Selye and his collaborators dealing with the effects of cold are summarized in his book *Stress*, and in the subsequent supplements to this book.

Amongst these changes in the rat are an initial fall in weight, followed by a period when the weight remains constant, which may be succeeded by steady growth. The suprarenals hypertrophy and there is a slow but steady rise in blood pressure. Dugal and his colleagues (12, 16a, b, c, 17, 18) have thoroughly investigated these changes and the rôle of ascorbic acid. The level of ascorbic acid in the tissues, particularly the liver, spleen, and testes, declines during the first few hours of exposure to cold. Thereafter the level gradually rises and continues to increase for many weeks (Fig. 54).

In the rat, this increase in the tissues is not due to a diminished

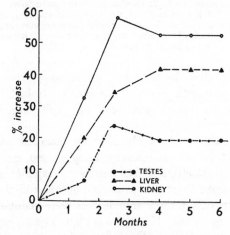

FIG. 54. The increase in ascorbic acid content of kidneys, liver and testes of rats kept in the cold. The percentage increase over controls is plotted on the ordinate against time spent in the cold in months on the abscissa. (From Dugal and Thérien, *Canad. J. Res.*, Sect. E., **25**, 1, 113, 1947.)

M

excretion, as the amount of ascorbic acid in the urine actually increases on exposure to cold (Fig. 55 (19a, b)). The rat synthesizes ascorbic acid, and does not normally require any in its diet. If, however, varying amounts are added to the rations, the changes on exposure to cold are modified. The survival rate is increased, and the degree of cold the rats can withstand is directly related to the dose of ascorbic acid. The initial loss of weight on exposure is

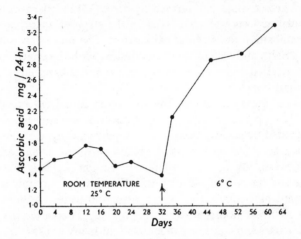

Fig. 55. The excretion of ascorbic acid of rats exposed to moderate cold (6°C (43°F)). The urinary excretion in mg/24 hours on the abscissa is plotted against time. At the point marked by an arrow the rats were moved from a comfortable environment (25°C (77°F)) to a cool one at 6°C (43°F). No ascorbic acid was given in the diet. There is an immediate and sustained increased excretion of ascorbic acid, accompanied by a slight weight loss. (Redrawn from Thérien and Dugal, *Rev. Canad. de biol.*, **8**, 3, 249, 1949.)

diminished, and after the period of stabilization there is a more rapid increase in weight as compared with controls. The level of ascorbic acid in the tissues rises less or not at all, according to the dose level.

The suprarenals undoubtedly play an essential part in the response to cold. In rats not receiving ascorbic acid, the content in the suprarenals is closely related to the weight response. In animals losing much weight there was a marked decrease in the ascorbic acid content, whereas in those animals who were maintaining their weight the content of ascorbic acid was high. The higher the content the less the weight loss. This relationship does not hold after the initial period of weight loss and stabilization. There is no correlation between the weight increase and ascorbic acid level in suprarenals. Similarly at normal temperatures, there is no such relationship (20).

There is a marked and sustained hypertrophy of the suprarenal gland on exposure to cold: this hypertrophy is prevented by the addition of ascorbic acid to the diet. It has been shown by Giroud (21) that ascorbic acid increases the formation and utilization of the cortical hormones. Sayers *et al.* (22) have shown that the cholesterol content of the suprarenal is a good index of its activity, so measurements were made in rats receiving ascorbic acid and controls with none, both in the cold and at normal room temperatures. A greater fall in the cholesterol content was observed in treated compared with non-treated animals after 24 hours' exposure, but after 72 hours there was no difference between the two groups. As a fall in the cholesterol content has been shown to be due to increased activity it can be concluded that ascorbic acid does not inhibit activity of the suprarenal cortex, although it does prevent hypertrophy (20).

Heroux and Dugal (23 a, b) also investigated the hypertension developing in rats during prolonged exposure to cold, and found that ascorbic acid diminished the hypertension when developed and prevented the rise in pressure when given throughout exposure. D.C.A. induced hypertension is also markedly diminished by ascorbic acid, so it was concluded that the effect on cold hypertension is due to a direct action on the adrenal cortex.

Dugal and his colleagues (24) have investigated other animals than rats, including guinea-pigs, rabbits and monkeys. In all animals investigated ascorbic acid increased the resistance to cold and diminished its damaging effects. Fig. 56 shows the result of an experiment on guinea-pigs.

In all these animals, including rats, ascorbic acid did not have any effect on the total calorie intake. It would appear that there is an increased requirement of ascorbic acid in the tissues, which may be concerned with the local utilization of cortical hormones. In animals able to synthesize, there is an increased formation of ascorbic acid, in other animals there is an increased requirement in the diet. No detailed metabolic investigations have as yet been made by Dugal and his colleagues. There is, however, an interesting series of investigations by Sellers and his colleagues (25, 26, 27, 28, 29, 30, 31) on the metabolism of rats exposed to moderate cold, without any added ascorbic acid. The metabolic rate is increased about $2\frac{1}{2}$ times, and there is an increased nitrogen excretion.

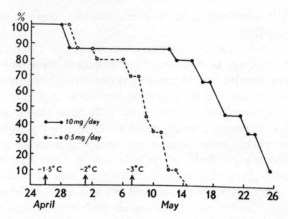

FIG. 56. The survival of rats on different doses of ascorbic acid, when exposed to cold. The percentage survival of the two groups is plotted on the ordinate against time in days on the abscissae. The arrows indicate the time at which the temperature of the cold room was changed. (From Dugal and Thérien, *Canad. J. Res.*, Sect. E, **25**, 1, 132, 1947.)

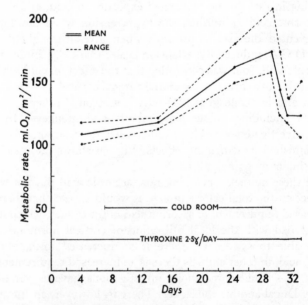

FIG. 57. The metabolic rate, measured at a room temperature of 30°C (86°F), of 4 thyroidectomized rats before, during and after exposure to cold (1·5°C (35°F)). The animals received 2·5 micrograms thyroxine/day. There is a considerable rise in the metabolic rate, although the animals are only receiving a maintenance dose of thyroxine. (From Sellers and You, *Am. J. Physiol.*, **163**, 81, 1950.)

The increased metabolism and nitrogen excretion develops rapidly, and this increase is obtained both in adrenalectomized and thyroidectomized animals (30). The fact that the initial rise in metabolism and protein catabolism is independent of the thyroid and suprarenal is not unexpected as the hyperplasia of these glands is relatively slow and takes at least several days to develop (Fig. 57).

Measurements of metabolic rates in animals exposed for long periods of time show that after the rapid initial rise the increased metabolism is maintained although after several months there may be a gradual decline. Sellers and You (26) measured the oxygen consumption both in the cold (M.R. 1·5°C (35°F)) and at a room temperature of 30°C (86°F) (M.R. 30°). M.R. 1·5° rose rapidly as described, whereas M.R. 30° increased only very gradually, but after 2 to 3 weeks was approximately 35 to 40 per cent higher than initial values (Fig. 58).

The problem that Sellers has investigated is the mechanism and significance of the increased metabolic rate. In other words, is

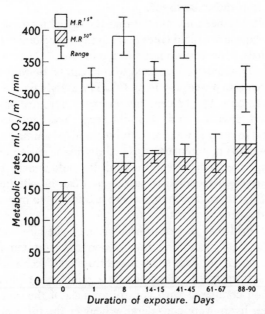

FIG. 58. The metabolic rate of rats after varying periods of exposure to cold (15°C (35°F)). The oxygen consumption in the cold rises rapidly, but does not increase significantly with continued exposure. There is a more gradual increase in the oxygen consumption measured at 30°C (86°F). (From Sellers and You, *Am. J. Physiol.*, **163**, 81, 1950.)

there any evidence of an extra muscular chemical regulation against cold developing during acclimatization, or can all the increase in metabolism be explained on the basis of an increased muscular activity? The fact that animals who have been living in the cold room show a gradual increase in metabolic rate when this is measured in a hot environment (30°C, 86°F) is suggestive evidence that non-muscular elements are involved. Sellers and You (26) attempted to eliminate muscular movements by anaesthetizing animals and measuring their oxygen consumption at room temperatures of 30°C (86°F). Control animals and cold acclimatized animals both showed drops of approximately 20 per cent, but the absolute values under anaesthesia were significantly higher in the cold acclimatized animals than in the controls. Although the authors are careful to point out that it was not possible to conclude that muscular relaxation was complete or necessarily similar in the two groups, yet the evidence is suggestive of increased metabolism in non-muscle tissue, on the assumption that anaesthesia depresses muscle activity, more than the metabolic activity of non-muscular tissue.

It has already been mentioned that the initial increase in metabolism on exposure to cold takes place in thyroidectomized animals. Such animals, however, do not survive for more than 24 hours in the cold, but animals acclimatized to cold can be thyroidectomized and survive up to 5 weeks at 1° to 2°C. The metabolic rate of these animals was only 10 per cent higher than thyroidectomized unacclimatized animals, and this difference was not statistically significant. On the other hand, thyroidectomized animals on a small (2·5 microgram) maintenance dose of thyroxine raised and maintained a high metabolic rate in the cold. It would appear, therefore, that the sustained metabolism in the cold can be obtained in the absence of the thyroid, although there is undoubted hyperplasia of the thyroid during acclimatization.

Further evidence of non-muscular, i.e. 'chemical' regulation was obtained by examining the oxygen consumption and succinoxidase activity of liver slices taken from cold acclimatized rats. The livers from these animals had a significantly higher oxygen consumption compared with control animals of the same weight, and as their livers were considerably heavier the inference is that the total metabolic rate of the liver in the acclimatized rats was approximately 50 per cent greater than the controls. The succinoxidase activity was also increased by about 50 per cent in the acclimatized animals. It is of interest in this connection that Malcolm Brown (32) found that in a considerable number of Eskimo, there was a great increase in the size of the liver. Biopsy specimens of these livers appeared to be normal.

The findings of Scholander and Irving described at the beginning of this chapter emphasized that in Arctic animals the main, indeed possibly the only, adaptation to cold was increased insulation. The effect of removal of fur by clipping was tested by Sellers, You and Thomas (27). Non-acclimatized animals only survived a few hours after clipping at a temperature of 1·5°C (34·7°F), whereas acclimatized animals were able to survive in the cold for many days or weeks. The metabolic rate of the non-acclimatized clipped rat rose to very high levels (430 ml. O_2/m²/min.); and in the cold acclimatized animals after clipping the metabolic rate was even higher (470 ml.O_2/m²/min.) and was maintained for over 20 days. The difference between the two groups was summarized by the authors in saying that the acclimatized rat can maintain a very high metabolic rate whereas the unacclimatized animal cannot, and therefore does not survive (29).

The onset of acclimatization and persistance of acclimatization was tested by clipping rats after varying periods in the cold room. Little difference was found in animals maintained for 1–14 days in the cold before clipping. After 3 weeks in the cold there was definite evidence of acclimatization as shown by the longer survival time after clipping, and when animals were kept 4–6 weeks in the cold apparently full acclimatization was developed. Animals which had been fully acclimatized were brought back to normal room temperature for 4 days, clipped and returned to the cold room. Their survival time was reduced and closely resembled the survival curves of 3-week adapted rats (Fig. 59). Acclimatization to cold appears to be fully developed in 4–6 weeks, and can be reduced by a short exposure to a warm environment (28).

It will be interesting to follow the future development of the important work so far carried out by the Quebec and Toronto groups. The integration of their separate lines of approach should prove fruitful. It is certain that in the rat acclimatization to cold can be developed and that acclimatization is enhanced by additional ascorbic acid. The essential feature of the acclimatization is an increased heat production (not affected by ascorbic acid) rather than an increased insulation, and there is strong evidence of increased heat production in tissues other than muscle. The

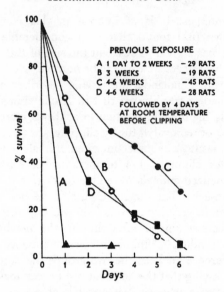

FIG. 59. The survival rate of clipped rats in the cold. The different groups had been exposed to cold for varying periods before clipping. The longer the period of previous exposure, the longer the survival time after clipping. The acclimatization to cold is greatly reduced by an intervening period at room temperature before clipping (group D). (From Sellers *et al.*, *Am. J. Physiol.*, **167**, 644, 1951.)

thyroid and the suprarenals hypertrophy during acclimatization, the latter effect being reduced or absent if ascorbic acid be given. However the increased metabolic rate can be observed under certain conditions in thyroidectomized animals, so the rôle of the thyroid is still obscure.

Cold acclimatization has been demonstrated in the rabbit and the rat by displaying a difference in susceptibility to cold injury (6). The lowest temperature to which these animals can be exposed without hypothermia or frostbite is −30°C (−22°F) for rabbits and −7°C (19·4°F) for rats. Animals were maintained at these respective temperatures for 16–20 hours per day for 7 weeks. Then the cold acclimatized rabbits with control animals were placed in a room at −45°C for 8 hours. All the controls suffered from frostbite and showed progressive hypothermia, whereas the cold acclimatized animals showed no ill effects at all. The same results were obtained with the cold acclimatized rats who were exposed to −15°C (5°F) for 5 hours. (See discussion in Macy Foundation Conference on Cold Injury, 1951.)

Adolph (33) has compared the response to hypothermia in control and cold-acclimatized rats. The latter were kept at 5°C (41°F) for 5–25 days, so presumably they were not fully acclimatized, although they showed some increase in the metabolic rate measured at 30°C (86°F). The cooling rates in these animals were not different from controls, and the lethal body temperature was not altered. The maximal oxygen consumption during hypothermia was also similar in control and acclimatized rats.

In these experiments partial acclimatization did not affect the response to hypothermia, but in a preliminary note by Adolph (34) it is stated that acclimatized rats exhibit slower cooling rates in moving air at 6°C (43°F) than controls.

Adolph (35, 36) has studied the responses to acute hypothermia in several species of varying ages, including cat, rat, mouse, guinea-pig and golden hamster. There is a striking feature in all these species, namely the increased tolerance to cold of the new-born animal. This might be thought to be related to the increased tolerance to anoxia which is also a feature of the infant animal, but Adolph has produced evidence to show that different mechanisms are involved.

It is of interest that Holtkamp, *et al.* (37) found that the adrenal cortical hormones do not protect infant rats, under 16 days of age, against the effect of cold. This is in contrast with the marked effect obtained in older animals with cortical hormones, and suggests that the change to the adult pattern found by Adolph at 18 days is accompanied by alterations in the response to these hormones.

The increase in metabolic rate in the cold has been demonstrated by many other workers. Collip and Billingsley (38) managed to eliminate shivering and any increase in muscle tonus by acclimatizing rats to low temperatures of −5°C (23°F). Previous to acclimatization metabolic rates at 16°C (61°F) were 433*l.* $O_2/m^2/$ day. After acclimatization to cold, rats curled up and went to sleep at 16°C (61°F), but the metabolic rate was virtually the same. This finding is consistent with chemical regulation but does not prove it; before acclimatization the high metabolic rate at 16°C could be attributed to increased muscle activity, after acclimatization the same rate might be achieved by increased metabolic rate of other tissues.

The effect of environmental temperature and metabolic rate has recently been studied again by Hart (39). As he points out, 'the energy expended to maintain the same body heat in homeotherms must increase when a decrease in environmental temperature occasions an increase in heat loss'. He was investigating the problem whether muscular work could supply the extra heat required.

The oxygen consumption of white mice was measured during rest and work at temperature levels between −9·6°C to 37°C (14·7° to 98·6°F). Cyclical variations were observed at all tem-

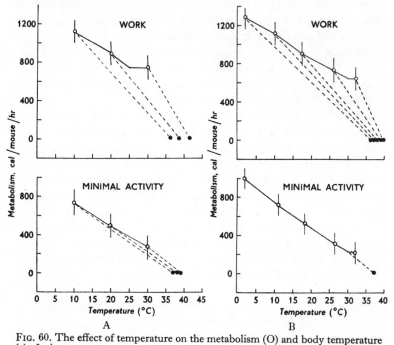

FIG. 60. The effect of temperature on the metabolism (O) and body temperature (●) of mice:—
A. Acclimatized to 20°C (68°F) and B. those acclimatized to 6°C (43°F).
Body insulation is represented by the slope of the dotted lines. There is a greater drop of body temperature in group A than in group B, on exposure to cold.
(From Hart, *Canad. J. Zool.*, **30**, 90, 1952.)

peratures, although the amplitude of the cycle was slightly greater at the higher temperature. Four levels of activity were observed, rest, light, moderate and maximum work. Animals at rest at −9·6°C (14·7°F) had a higher oxygen consumption than animals working maximally at 32°C (89·6°F). From 32°C (89·6°F) to −9·6°C (14·7°F) there was a steady increase in metabolic rate from 50 to 250 ml./O_2/hr. Temperature and work produced additive effects on metabolism as long as work was below maximum (Fig. 60). This shows that muscular exercise cannot substitute for the metabolic response to cold.

There are many other important studies bearing on the acclimatization of animals to cold, but these cannot be reviewed in detail, as the emphasis in this work is on studies on man. The results of these animal experiments can be summarized in order to clarify the subject and to indicate the features which may be examined in man.

Naturally occurring adaptation to cold is essentially due to increased insulation. The insulation is provided by fur in land animals and by fat in aquatic animals. Considerable variation of the insulation is achieved by degrees of pilo-erection, by variations of superficial blood flow and by posture. Increased metabolism or changes in body temperature appear to be unimportant in Arctic animals.

Acclimatization to cold can be induced in laboratory animals, and can be demonstrated by increases in survival time in severe cold, with absence or diminution of the incidence of cold injury as compared with control animals. There is a rapid increase in metabolic rate in the cold, with a more gradual increase in 'basal' metabolic rate, on prolonged exposure. Induced acclimatization in some species, e.g. rat, is due to increased heat production, although increased insulation cannot be entirely ruled out. The increased metabolic rate is accompanied by an increased food intake, although there may be a decrease in body weight or a check on the growth of young animals, at any rate initially.

The adrenal cortex and the thyroid gland both hypertrophy in animals exposed to cold for moderately long periods. The hypertrophy of the adrenal gland and survival in the cold is affected by the quantity of ascorbic acid in the diet: the lower the temperature the greater the quantity of ascorbic acid required.

Acclimatization in Man

Examination of the response of man to prolonged exposure to cold, in a search for evidence of acclimatization, should include measurements of (*a*) heat production, (*b*) calorie intake, (*c*) evidence of hormonal changes, (*d*) changes in body insulation, including thickness of body fat, (*e*) changes in vasomotor control of superficial vasculature, (*f*) changes in tolerance and performance, (*g*) evidence of diminished or increased incidence of cold injury, (*h*) evidence of local acclimatization, (*i*) duration and persistence of any changes detected.

Clear evidence is not available on all these points; on the other hand, measurements have been made on man which do not appear to have been attempted on animals, e.g. changes in blood volume and in the peripheral circulation.

Basal Metabolic Rate and Diet in the Cold

The metabolic rate is increased in cold environments, both in man and laboratory animals. The basal metabolic rates of Eskimo have been measured by several groups of workers, who have found increases. Krogh and Krogh (40) were the earliest to study metabolism in the Eskimo with accuracy, and they reported a raised B.M.R. The following increases have been recorded:

	B.M.R. %
Heinbecker (41)	+ 33
Crile and Quiring (42)	+ 14·5 in male, + 16·4 in female
Rabinowitch *et al.* (43)	+ 26
Høygaard (44)	+ 13

However, Levine (45, 46) found that the basal rates of Eskimos at Point Barrow, Alaska, were similar to white people in the same area. His studies were carried out in the summer months. Recently, Bollerud, Edwards and Blakely (47) have re-investigated the problem. Their measurements were made in 1949–50, during the winter, on 10 male Eskimos in 1949 and 23 male Eskimos in 1950. Comparison was made with a group of Air Force personnel living in the same area.

%
The 1949 Eskimos had an average B.M.R. of + 17
The 1950 Eskimos had an average B.M.R. of + 14
The control group had an average B.M.R. of + 8·5

At the time of the test the Eskimos were eating a high fat, high protein diet. They were fasted for 12 hours before metabolic rates were determined. The control group were consuming approximately 3,000 calories per day.

Various criticisms have been made of the estimates of B.M.R. in Eskimos. These are mainly concerned with the metabolic state of the Eskimo at the time of the determination and the possibility that he was not in a basal state.

Queen's University, Kingston, Ontario (48), has sent an expedition to Southampton Island for several successive years to study the Eskimo. Such studies have included considerable medical services, and as a result the members of the expedition are regarded as benefactors, and co-operation is extremely good. Basal metabolic rates were estimated in the subject's hut or tent and while he was in his sleeping bag. On most occasions the subject slept during the determination.

Sixteen subjects, who were all considered to be free from disease, were studied. Four series of observations were made, starting at the beginning of the summer, i.e. as the last of the snow on the ground was melting: and subsequent measurements were made at two-week intervals. The results showed a substantial increase in B.M.R.

A	B	C	D
+ 31·4	+ 27·6	+ 27·2	+ 23·7
July 13th	July 27th	August 10th	August 24th

The subjects were familiar with the equipment and the investigators: they were not disturbed by the measurement, so it is reasonable to assume that these are correct basal metabolic rates. They tend to show a steady decline during the summer: the difference between *A–D* is statistically significant.

Heinbecker (49) has produced evidence that the elevated metabolism is probably the direct result of a high protein dietary and not of cold *per se*. In rats, the rise of B.M.R. in the cold was only found when the calorie intake was increased as calorie requirement rose. On the other hand, as pointed out above, Sellers found an increased metabolic rate when rats were on the same intake in cold as in heat. He also found that allowing an increased intake raised the metabolic rate further.

It is difficult to estimate the calorie intake of Eskimos in the Arctic, as there is a very marked variation from day to day and indeed from meal to meal (48). The average intake of walrus meat in one group was 343 grams (range 150–680 grams), 114 grams of fat (range 35–245 grams) and 208 grams of bannock (range 0–430 grams). When caribou meat was provided the ranges were, meat 570–1,110 grams, 0–173 grams of fat, and 140–200 grams of bannock. The average intake probably represents about 2,500 calories/day in the walrus meat period, fat representing about 1,000 calories, and in a second period 3,250 calories, of which some 2,700 calories came from fat. In a third period, the intake averaged 2,600 calories with only 350 calories from fat. Each period lasted 3 days, and the group consisted of 9 members of two families, 2 being children. Observations on an adult, weight 144 lb. over a period of 6 days, are illustrated in Fig. 61. There are the most remarkable variations in the proportion of meat and fat and a range

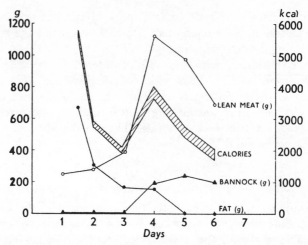

F IG. 61. Daily dietary of an Eskimo for one week. The right ordinate gives the calorie value of the diet, and the left hand the weight in grams of the various foods. The calories value is given as a hatched band representing the probable upper and lower limits, calculated from the diet. It will be seen that there is a very marked daily variation in total calorie expenditure, and in the components of the diet.

of calories/diem from 2,000–5,700. A detailed study of the dietary of the Greenland Eskimo has been published by Høygaard (44). They illustrate the great difficulty of obtaining representative values. When the complete reports are published they should be consulted for further details.

It has been shown in laboratory studies on rats that a high fat diet promotes survival in the cold (12). However, the evidence from other studies is not clear, except that protein does not protect against cold, but fat and carbohydrate are possibly of equal value (50, 51).

There are many reports on the rôle of fat in the diet in Arctic conditions, which in general indicate (*a*) that an increased fat content is preferred, (*b*) that high fat content diets are not harmful (52, 53). Frazier (54) stated that 'individuals (members of the U.S. Antarctic Expedition 1939–1941) who had an abhorrence of fats at home, would eat butter or fat meals in great quantity'.

Keeton *et al.* (55) compared the effect of diets, of varying composition, on cold tolerance in a group of young healthy volunteers. A high carbohydrate diet was significantly more effective than a high protein diet, as regards maintenance of thermal balance and in the performance of various psycho-motor tests.

Further investigations were carried out by Mitchell *et al.* (56) in which it was found that a high fat diet was superior to a high carbohydrate diet only when meals were given at 2-hour intervals during the period of cold exposure (8 hours). When only one meal was given, there was no significant difference.

The interesting suggestion is made that this effect of frequent high fat meals may be due to a temporary deposition of dietary fat in the subdermal tissues following a high fat meal. Such a deposition would diminish the thermal conductivity of the superficial tissues and so reduce heat loss. Mitchell *et al.* point out that Schoenheimer and Rittenberg (57) demonstrated that the bulk of the dietary fat, in the mouse, is conveyed to the fat depots before utilization.

The same group investigated the effects of varying the intake of thiamin, riboflavin, niacin and ascorbic acid, over a period of 3 months (58). No effect could be demonstrated on the maintenance of heat balance, or on the performance of psycho-motor tests during cold exposure. There was a slight increase in the excretion of nicotinic acid and a decrease in ascorbic acid output. Similarly, Blair (59) was unable to demonstrate any benefit from vitamin supplements on subjects living at Fort Churchill.

The specific dynamic action of food (S.D.A.) might well be thought to be of considerable importance in the protection afforded by various diets against the effects of cold. However, Glickman *et al.* analysing the effect point out that even with a high protein meal, when the peak S.D.A. reaches 33 cal/hr above basal, the effect is trivial. As it has been clearly demonstrated that high protein meals are of significantly less effect than high fat or carbohydrate with lower S.D.A. it would appear that the S.D.A. is of little importance as regards protection against cold.

The work on nutrition has been excellently reviewed by Mitchell and Edman (60), and considerable use has been made of their paper in preparing the section above.

Summary

Nutritional studies on man exposed to cold demonstrate an increased calorie intake, which is in part due to the extra metabolic cost of activities caused by the hampering action and weight of Arctic clothing. A high fat diet may be more effective in providing protection against heat loss than a high carbohydrate diet. A high protein diet is definitely less effective.

There is no satisfactory evidence so far that there is a greater requirement of any vitamin so far as man is concerned. Studies which are in progress will, it is hoped, give much needed information on the rôle, if any, of ascorbic acid, in protecting man against the cold.

Tolerance to Cold

The next aspect of acclimatization to cold to be considered concerns the changes which have been observed affecting tolerance to cold. Such tolerance can be estimated by measuring the time elapsing after exposure to a given temperature at which shivering commences, or when the subject complains of discomfort of varying degree.

Ames *et al.* (61) in their investigations on methods of rewarming following exposure to cold, observed that after repeated trials on the same subjects there was evidence of an improved tolerance. Four subjects were exposed on 16 occasions in the course of 3 weeks to a temperature of $-40°C$ ($-40°F$) for 3 hours a day. They wore Arctic clothing providing approximately 3·4 clo of insulation. The subjects sat for the first hour. During the second hour they tried various rewarming techniques and sat quietly for the third hour. The first, sixth and sixteenth days were used as control periods, during which the subjects sat quietly for two hours at $-40°C$ ($-40°F$). The time which elapsed before subjects complained of being uncomfortably cold increased from an average of $72\frac{1}{2}$ minutes to 90 minutes, comparing the first and sixteenth day. There was considerable variation between subjects, however, and these differences are not statistically significant. The onset of shivering was also delayed, coming on at $79\frac{1}{2}$ minutes on the average on the first day and only after 109 minutes on the last day. This difference is significant. When objective data were used only slight differences could be demonstrated. There was a small increase, which was significant, in the heat content of the body after 2 hours' exposure on the last day compared with the first, but there were no significant differences in rectal temperature or toe temperature. Average skin temperature was some $1·1°C$ ($2°F$), (Fig. 23) higher at the end of exposure on the last day than it had been on the first occasion, but this was not significant. No correlation was possible with changes in metabolic rate as there were wide fluctuations in any one subject from day to day.

Glickman *et al.* (58), who were studying diet in relationship to cold tolerance, exposed their subjects to various temperatures with different clothing. They obtained evidence suggestive of acclimatization after a considerable period of daily exposure, as the rectal temperature fall in the cold was significantly smaller. Their subjects were either in a cool environment, 15·5°C (60°F), in which they only wore a union suit, or in a cold room, −29°C (−20°F) in Arctic clothing. Eight hours daily were spent in these rooms. In one group during the initial period of exposure of 25 days in the cool room, the average drop of rectal temperature was 0·71°C (1·28°F) in the 8 hours. A further 79 days with intervals was spent in the cold room, and a final exposure of 7 days in the cool. The average drop in rectal temperature in the last 7 days was only 0·2°C (0·36°F). This acclimatization was completely extinguished after 34 days in temperate conditions, and was almost abolished at the end of 17 days.

On the other hand, Horvath, Freedman and Golden (62), who have carried out many experiments in low temperature chambers, have been unable to obtain convincing evidence of acclimatization to cold (Horvath *et al.* (63 a, b, c, d)). Similarly, Adolph and Molnar (64) did not find any changes in their subjects which could be adduced in support of physiological acclimatization. Details of these experiments are given below.

Stein, Elliott and Bader (65, a b) examined the reactions of a group of healthy young male adults who were exposed for 5 hours daily either in a hot room or a cold room. The object of the experiment was to study adrenal and thyroid responses, and to determine the effect of acclimatization to heat on the responses in the cold and vice versa.

The three subjects spent five hours in the hot room (D.B. 41·5°C (107°F), W.B. 31·5°C (89°F), air 3 m.p.h.) on five days in the week with a total of 19 exposures; this was immediately followed by fourteen 5-hour periods in the cold room (−29°C (−20°F), air 3–4 m.p.h.) and there were then five re-exposures to heat. The subjects had a 5-week interval with no environmental stress, and then experienced three periods in the hot room. Heavy Arctic uniform was worn in the cold room, and exercise was taken by walking on the treadmill at 3 m.p.h. Whenever the toe temperature fell below 7°C (45°F) exercise was started and continued until the feet had warmed and toe temperature was over 10°C (50°F).

Heat acclimatization was fully developed, as shown by the

N

diminishing effect of exercise on pulse rate, rectal temperature, metabolic rate and weight loss, after 10 exposures to heat. There was scarcely any loss of this acclimatization after the 14 days of cold exposure, but following the rest of 5 weeks, on the first re-exposure to heat a marked diminution of acclimatization was apparent. During the period of cold stress, it was found that the rate of drop of toe temperature increased, and this was interpreted to mean that there was a more rapid and intense peripheral vasoconstriction as cold acclimatization developed. This finding is in contrast to the report of Ames *et al.* which has already been discussed, in which toe temperature after 2 hours in the cold showed a slight tendency to be higher with repeated exposures. Rectal temperature or skin temperatures other than the toe did not change, nor was there any difference in the time of onset of shivering. There was some haemo-concentration during the cold exposure, but no significant change in blood volume.

There was a considerable diuresis on exposure to cold which persisted for the entire period with increased chloride excretion, so the subjects were in negative chloride and water balance. However, the authors were unable to find any clear evidence of increased tolerance to cold.

As evidence of the stress response to cold, it was found that there was a short decline in eosinophils during the cold exposure. The A.C.T.H. response was normal in two of the three subjects after the period in the cold, but was reduced in the third subject. It was not possible to conclude that there was any significant decrease in the adrenal cortical reserve as a result of either the heat or cold exposure. The basal metabolic rate was not significantly altered in the 14 days of cold exposure.

The studies of Bazett and his colleagues on acclimatization (66, 67, 68) are of classic importance. Their results are so well known that it is unnecessary to give a detailed account, and only a brief summary is included: the original papers should be carefully read by all those interested in this field. Their subjects were exposed continuously in a hot room environment 32°C (89·6°F) and then changed to a cool room. It was found that the blood volume gradually increased in the heat and conversely decreased in the cold. Haemo-concentration occurred initially, but as well as loss of plasma there also appeared to be an absolute loss of haemoglobin from the circulation.

On changing from the heat to the cold, there was an immediate increase in the calorie intake, of the order of 14 per cent. The rise in food consumed was not accompanied by a rise in metabolic expenditure. On the contrary these workers found an initial decline in basal metabolic rate which persisted for 2–3 days and then rose progressively. The fall in plasma volume was accompanied by a large diuresis which persisted for some time, and by a considerable loss of chloride. These results are confirmed as regards the diuresis and chloride loss by Stein *et al.* (65 a). The effects of cold on metabolic rate are far from clear: Burton *et al.* (67) reported that in spite of the diuresis in the cold there was a relative or even absolute weight gain, which would support their finding of an increased calorie intake compared with expenditure. It will be recalled that in animals there is an immediate increase in total metabolic rate, but a very slow rise in the basal values. The increased food intake appears to be an appetite effect: it is a matter of common observation that appetites are very great in the north land of Canada, especially amongst newcomers, even though they are not unduly exposed to cold or exertion.

The circulatory changes in the cold included a rise in cardiac output, an increased blood pressure and a higher peripheral resistance. The changes in cardiac output on standing were different in the heat from the cold. There was little or no reduction in output on standing in the warmth but a marked drop in the cold room. The blood flow in the fingers in the first 18 hours in the cold was higher than the final values, and full constriction only developed on the second or third day (Fig. 62). The forearm veins did not show full constriction either until the subject had been in the cold room for 2–3 days.

Bazett (69) has put forward the view that in hot conditions the venous return is mainly in the superficial veins but in the cold the superficial veins constrict and there is a relatively greater flow in the deep venae comites.

There is an important difference between the work of Bazett *et al.* and most of the other studies on acclimatization. The exposure was continuous in Bazett's experiments and the level of temperature was relatively high. Much of the work that has been done in other laboratories has consisted of short exposures (2–5 hours) at very low temperatures, repeated on many occasions. There is no doubt that acclimatization to heat can be developed by daily periods of 4 hours in a hot room, the rest of the time being spent in temperate conditions. It is possible that short daily exposures

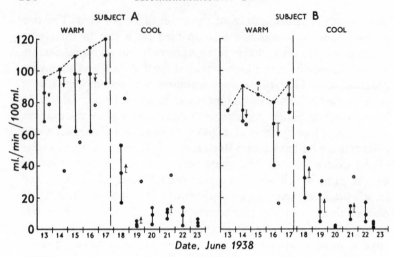

FIG. 62. Blood flow through the fingers (ordinates) during successive days (abscissa) of exposure to a warm and a cool environment. Average, maximal and minimal values of blood flow are indicated. The open circles indicate the blood flow when the legs were heated, the arrows indicating the direction of change from the average values. Vasoconstriction in the finger develops relatively slowly in the cold. (From Scott *et al., Am. J. Physiol.,* **129,** 102, 1940.)

may not be adequate to develop full acclimatization to cold except after a long period.

Horvath *et al.* (62) kept their subjects continuously in a cold room as low as −29°C (−20°F) for 8 days. There was no initial drop in metabolic rate under these conditions; instead oxygen consumption measured in the sitting positions rose immediately by 30 per cent and stayed approximately constant during the rest of the experiment. When the subjects returned to an environment of 25°C (77°F), the metabolic rate was significantly higher than control values. There was considerable individual variation between the subjects in other responses. During periods of rest, toe temperatures fell, but the rate of fall progressively decreased in one subject suggesting acclimatization. In another subject the rate of fall was remarkably constant throughout the experimental period. Three of the four subjects failed to report any subjective improvement, while the remaining one had definite increased tolerance to cold. Horvath *et al.* did not consider that they had succeeded in demonstrating any significant acclimatization to cold in the 8 days of the trial.

Adolph and Molnar (64) measured heat exchanges in men who

spent 4 hours a day, dressed in shorts only, out of doors at temperatures ranging from 0° to 31°C (32° to 88°F), each subject being exposed approximately 60 times. Their important paper details the responses to varying conditions, but acclimatization to cold was not observed. Their conclusions may be quoted: 'There is no evidence that skin temperature changed its relation to air temperature or that rectal temperature was either more or less constant. Shivering appeared under the same conditions after as before the various exposures, and oxygen consumption accelerated no more readily. The mental outlook may have changed slightly as apprehensions were dispelled by successive exposures, but even of that there was no evidence. It is true that in general the most severe exposures came late in the season, but there were numerous reversals in the order with which the various air temperatures were represented to the subjects. While one subject thought he tolerated cold better at the end of the season, another stated that he felt more sensitive to cold. Hence nothing could be concluded with respect to acclimatization.'

Glaser (70 a, b, c) studied the effects of exposing subjects for three successive 72-hour periods alternately in a cold, $-1°$ to $+ 3°C$ (30° to 37·4°F) and a hot room, 35°C (95°F). All subjects noted improved tolerance by the third day of exposure, and the second period of exposure to a cold room was always considered less uncomfortable than the first. A small decrease in the vital capacity amounting to an average of 200 ml. was obtained in the cold room, confirming observations made during acute exposures to cold. There was also a significant reduction in the volume of hand and forearm in the cold compared to the heat, amounting to an average of 60 ml. Scott *et al.* (68) had noted a small increase in forearm volume in the heat and decrease in the cold, which developed gradually over several days. Skin temperature and rectal temperature were both significantly lower in the cold room, as was the pulse rate. Glaser also found an increased haematocrit in the cold as so many workers have done. On the third day in the cold room there was an average rise of $0·65°C$ ($1·17°F$) in skin temperature, compared with the first day and a significant rise of rectal temperature amounting to $0·32°C$ ($0·58°F$). There was a greater rise in skin temperature of $1·1°C$ ($2·0°F$) comparing the second with the first exposure. These results indicate that continuous exposure to moderate cold induces acclimatization.

A recent report by Carlson *et al.* (71) also gives evidence of

acclimatization to cold. A number of subjects were examined in
Alaska, during three periods of five days in which several hours
were spent out of doors at temperatures of $-23°$ to $-26°C$ ($-10°$
to $-15°F$). There were three groups, one of which consisted of
the authors who travelled to Alaska for the experiment and were
therefore considered non-acclimatized. A second group was termed
partially acclimatized as the members had spent some time in
Alaska. The third group included two Eskimos and two Americans
who had lived in Alaska, working out of doors for several hours
each day. This group was called fully acclimatized.

There was an increased calorie intake by the non-acclimatized
group amounting to an average of 25 per cent when cold exposure
began. No difference was observed in the rate of heat loss between
the different groups in the cold. However there were clear indications of differences between the groups in the mode of heat loss.
In the words of the report 'the acclimatized man maintains a relatively lower level of metabolism, in the cold, provides a smaller
percentage of the total heat loss with metabolism, and donates a
larger proportion of his body weight to losing heat. Thus it can be
concluded that the acclimatized man maintains a smaller core, so
that under conditions of cold stress he has a smaller portion of his
body to maintain at constant maximum temperature and can, in
effect, store up more heat during periods of activity at times of no
cold stress, to be used when and if cold stress increases. Under continued stress this thicker shell allows the acclimatized man to lose
more stored heat than the non-acclimatized before his metabolism
must increase. By this subtle reduction of body core, the acclimatized individual gains several advantages. He can stay out under
cold stress for longer periods of time, and can drop the total heat
load to a lower level before he must start shivering or retire to a
warmer place or put on warmer clothing, and his level of exercise
to keep him from shivering will be lower. He will be able to maintain circulation in the extremities longer with consequent increased
dexterity and protection against frostbite.'

The authors also found that the skin temperature of the finger in
the acclimatized subjects was higher at the beginning of cold stress,
and dropped less than in the non-acclimatized subjects. These
results confirm those of Balke, *et al.* (72), that there is an increased
peripheral blood flow with acclimatization.

Carlson *et al.* (71) included subjects who were permanent inhabitants of the Arctic. If acclimatization to cold does exist in

man, it should certainly be demonstrated more easily in those who spend very long periods in Arctic conditions. It is therefore of particular interest to record the findings of observers on Antarctic or Arctic expeditions. Unfortunately, the difficulties involved in making accurate measurements under the conditions existing in such expeditions are so considerable that there is a relative paucity of reliable information. There are many indications in the reports of such expeditions which would support the view that acclimatization to cold, in the sense of increased tolerance, gradually develops. However, detailed reference will be made only to two of these contributions, as it is not considered justifiable to include material which is only subjective. Frazier (54) was the medical officer of the U.S. Antarctic Service Expedition, and spent a year in Little America III. He observed that during the initial period in the South all available clothing was worn, but as winter approached and temperatures dropped, no further protection was sought. In fact many of the members of the expedition wore less during the winter than during the initial period. Some of the expedition spent the greater part of their time indoors owing to the nature of their work, whereas others were out of doors in very severe weather on sledging expeditions for long periods. When two such groups were compared by exposing them out of doors at low temperatures in the wind, frostbite appeared on the face within 20 to 90 sec. in the indoor group, whereas the freezing time of the outdoor group was approximately ten minutes. Frazier noted the dryness of the skin in the outdoor group and considered that cutaneous dehydration was partially responsible for this delayed frostbite. He also noted the avidity for fat amongst all members of the expedition.

Butson (73) who spent a year in the Antarctic also stressed that as the weather became colder, little or no additional clothing was worn. During the winter there were rare occasions when the temperature rose from $-29°$ to $0°C$ ($-20°$ to $+32°F$) overnight, and during such spells there were considerable complaints of the discomfort of the 'heat wave'. Butson found a small rise in B.M.R. after some months in the Antarctic of the order of 5 per cent. Amongst his other findings may be mentioned a rise in the fasting blood sugar, also noted by Frazier, and a bigger cold pressor response in those who suffered most from cold hands.

Local Acclimatization

There have been several reports suggestive of local acclimatization to cold. Manning (1) states that the Eskimos have a far greater tolerance to cold in the hands than any white man. Belding (74) found that the cold pressor test was negative in Nova Scotia fishermen whose hands were continually in very cold water. Mackworth (75) has, however, contributed the clearest evidence of local acclimatization. His technique was very simple and consisted essentially of measurements of tactile discrimination in the finger-tip before, during and after local cold exposure.

The first investigation was carried out at Fort Churchill. The subjects, who wore adequate clothing, exposed one finger to air at temperatures ranging from −25° to −35°C (−13° to −31°F), with air movements of 1–10 m.p.h. Tactile discrimination decreased during exposure, which lasted for 3 minutes, and then recovered when the subject returned to shelter. Results obtained under varying conditions were classified into four main groups.

	Calm air 0–5 m.p.h.	*Breeze* 6–10 m.p.h.
Cold	−25° to −30°C (−13° to −22°F)	−25° to −30°C (−13° to −22°F)
Very cold	−30° to −35°C (−22° to −31°F)	−30° to −35°C (−22° to −31°F)

Thirty-five volunteers were studied, of whom 9 had spent 1–2 years in the Arctic and the rest had been there less than 1 year. The results obtained are shown in Fig. 63 and emphasize once more the marked effect of air movement. Mackworth then divided his subjects into outdoor and indoor workers: the results are shown in Fig. 64. The conditions of this experiment were wind 4 m.p.h., air temperature −30°C (−22°F). There was a highly significant difference in the numbness, which was much greater in the indoor workers, who were only occasionally exposed to cold conditions. It should be mentioned that indoor temperatures at Fort Churchill are high, of the order of 24°–27°C (75°–80°F). The acclimatization effect, however, breaks down in more severe conditions. Mackworth points out that there might have been an inherent difference initially between the two groups, a degree of self selection. The outdoor workers, being familiar with cold, could also have been less apprehensive of cold exposure, and fear can produce vasoconstriction in the fingers.

A further set of experiments under carefully controlled conditions were then performed in a cold room at Cambridge. Fourteen subjects spent 2 hours a day 5 days a week for 5 weeks at −15°C

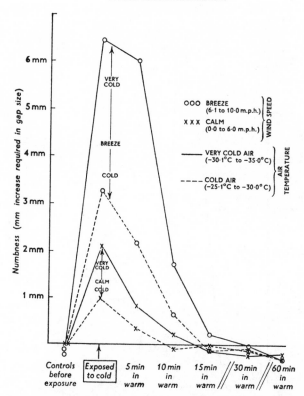

FIG. 63. The effect of air movement at different air temperature on the degree of numbness (ordinate) in fingers exposed for 1 minute. The effect of air movement is greater than the effect of temperature. (From Mackworth, *J. Appl. Physiol.*, **5**, 9, 533, 1953.)

(5°F). Arctic clothing was not worn, the men were dressed in Naval rig, including thick serge trousers and sweater. No gloves were worn. The subjects sat quietly for two hours and then the test finger was exposed to a 6 m.p.h. wind, also at −15°C (5°F) for 1 minute. The loss of tactile discrimination measured after the exposure steadily declined for the first 4 weeks and reached a steady level significantly lower than the initial results, during the 5th week.

Another experiment was carried out during the summer when outdoor temperatures averaged approximately 21°C (70°F). No downward trend was observed in a 4-week trial.

Two important conclusions can be drawn from these experiments. Firstly, local acclimatization to cold can be developed in

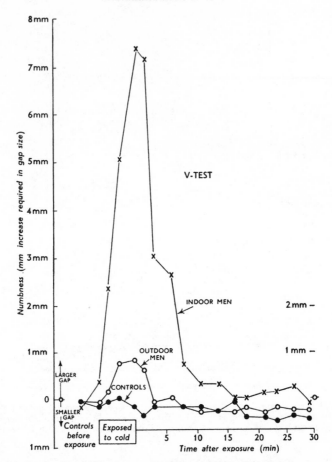

FIG. 64. Comparison of the degree of numbness (ordinate) in fingers exposed to cold air, in outdoor and indoor workers at Fort Churchill. There is a much greater effect of cold on the indoor than on the outdoor subjects (acclimatized). (From Mackworth, *Tr. J. Inst. Marine Engineers*, **LXIV** 6, 5. 1952.)

the finger. Secondly, exposure to heat can prevent this acclimatization. The mechanism of the local changes has still to be investigated, but it is probable from the evidence presented both by Mackworth and others that it has a vascular basis. In addition there may be an alteration in the skin with diminished water content and diminished water loss from the skin. The theory that vascular changes may be the basis of local acclimatization, receives support from studies on the hand blood flow in the Eskimo, recently reported by Brown and Page (76). Measurements were made with an air-filled plethysmograph at a room temperature of 20°C (68°F), and were compared with results obtained on white medical students. The skin temperature and blood flow in the hand were both considerably higher in the Eskimo than the controls. Higher hand flows were also recorded in the Eskimo when the hand and forearm was immersed on different occasions in water at temperatures ranging from 5° to 42·5°C (41° to 108·5°F). It was found that the hand blood flow did not decline so abruptly in cold water, as it did in the control group. The authors suggest that the well-known ability of Eskimos to work better than white men when their hands are bared in the cold may be partly accounted for by the higher sustained rate of blood flow and the more gradual initial decline. It is interesting that they also found that the clothing, which was similar for the groups, was comfortable for the Eskimo, but was considered inadequate by the students.

Mitchell *et al.* (56) considered the possibility that high fat diets protect against cold because fat is deposited in the subcutaneous layers before utilization. Some very interesting experiments have been carried out by Schmidt-Nielsen (77 a, b) on the different melting points of human fat, taken from varying sites. Visceral fat and subcutaneous fat from the abdomen had melting points between 32° to 36°C (89·6° to 96·8°F), whereas fat from the hand had melting points between 20°–27°C (68 to 80·7°F). Much lower melting points were found in fat on the foot and the ankle. A suggestive finding was that the melting point of the fat in one case was very high, even in fat removed from the extremities: the body was that of a stoker, who was habitually exposed to very high temperatures. Scholander *et al.* (4 c) quote unpublished results which show that the melting point of caribou fat progressively declines the farther down the leg it is taken. Acclimatization may therefore be due in part or accompanied by a physical change in the composition of the tissues.

This brief review of acclimatization in man may well be confusing. Certainly there are many contradictory results by different workers, and many completely negative reports. A few tentative conclusions appear to be justified. The evidence is convincing regarding the desirable composition of the diet in the cold, in that a high protein diet is unsatisfactory. There is important evidence from laboratory work showing that a high fat diet provides more protection than a high carbohydrate diet. This finding supports the many reports of Arctic explorers and those who live in the North that fat is desired in cold weather, and there appears to be 'fat hunger'. In spite of the powerful evidence from animal work, the rôle of ascorbic acid in protection against cold in man is still dubious.

The increased metabolism in the cold is largely explained by increased muscular activity, both voluntary and by shivering. There is also the increased metabolic cost of work, owing to the weight and hindrance of Arctic clothing. There remains the problem of 'chemical' regulation or increased metabolism by other tissues, and whether this plays an important rôle in acclimatization.

It will be remembered that the increased 'basal' metabolism found in laboratory animals in the cold developed very slowly and only with continued exposure. Virtually all laboratory experiments on man have been, by comparison, very short indeed. The negative results that have usually been obtained cannot be considered as showing evidence against extra-muscular 'chemical' regulation. There is some valuable positive evidence, e.g. Horvath found that increased metabolism, measured in the cold, persisted when subjects were returned to warm rooms.

The most important results are the measurements made on permanent inhabitants in the North, e.g. the Eskimo. Criticism can undoubtedly be levelled at some of the reported results, but measurements show an increased B.M.R. This may be evidence of a racial difference, i.e. adaptation rather than acclimatization. It is of very considerable importance to obtain clear evidence on this point, and it is hoped that due provision will be made in all future long-term expeditions to high latitudes for such determinations to be made on others than the Eskimo.

Evidence of acclimatization to cold has been sought in a number of ways, including subjective and objective. Several workers report increased tolerance times and delayed onset of shivering after repeated exposures (61). Others have failed to obtain such results (62). There have been considerable differences in the experimental conditions, and in view of Mackworth's finding that definite local acclimatization to cold can only be demonstrated with moderate exposure and vanishes under severe stress, further experiments under a variety of conditions are highly necessary.

Objective evidence of acclimatization includes diminished falls in body temperature and/or of skin temperature. Glickman *et al.* (58) and Glaser (70) have obtained significant changes in this respect. The results reported by Carlson *et al.* (71) strongly suggest an altered mechanism of heat loss in those habituated to cold.

There is also considerable agreement concerning the vascular changes produced by exposure to cold. There is an initial haemoconcentration, with a diminution of blood volume, including a loss both of plasma and red cells. There is a marked diuresis in the cold which persists for some time, and may be associated with the fall in plasma volume.

It seems reasonable to sum up in favour of the positive evidence that acclimatization to cold can under certain conditions be demonstrated in man. Much work will be required to evaluate these conditions and to lay down criteria before a clear verdict can be obtained.

REFERENCES

1. Defence Research Board Conference on Cold. Kingston, Ontario. December, 1950.
2. FRY, F. E. J. Effects of Environment on Animal Activity. Publications of the Ontario Fish Research Laboratories, 1947.
3. YAGLOU, C. P. Comfort Zone for Men at Rest and Stripped to Waist. *J. Indust. Hyg.*, **8,** 5, 1926.
4. (a) SCHOLANDER, P. F., WALTERS, V., HOCK, R. and IRVING, L. Body Insulation of some Arctic and Tropical Mammals and Birds. *Biol. Bull.*, **99,** 225, 1950.
   (b) SCHOLANDER, P. F., HOCK, R., WALTERS, V., JOHNSON, F. and IRVING, L. Heat Regulation in some Arctic and Tropical Mammals and Birds. *Biol. Bull.*, **99,** 237, 1950.
   (c) SCHOLANDER, P. F., HOCK, R., WALTERS, V. and IRVING, L. Adaptation to Cold in Arctic and Tropical Mammals and Birds in Relation to Body Temperature, Insulation and B.M.R. *Biol. Bull.*, **99,** 259, 1950.
5. IRVING, L. Physiological Adaptation to Cold in Arctic and Tropic Mammals. *Fed. Proc.*, **10,** 543, 1951.
6. BLAIR, J. R., DIMITROFF, J. M. and HINGELEY, J. E. Acquired Resistance to Cold in Rabbit and Rat. *Fed. Proc.*, **10,** 1951.
7. CHINN, H. I., OBERST, F. W., BYMAN, B. and FENTON, K. Biochemical Changes in Rats Exposed to Cold. U.S.A.F. School of Aviation Medicine, Randolph Field, Sept. 1950, No. 21–23–027.
8. STARR, P. and ROSKELLEY, R. A Comparison of the Effects of Cold and Thyrotropic Hormone on the Thyroid Gland. *Am. J. Physiol.*, **130,** 549, 1940.
9. (a) LEBLOND, G. P. and GROSS, J. Thyroidectomy on Resistance to Low Environmental Temperature. *Endocrinology*, **33,** 155, 1943.
   (b) LEBLOND, G. P., GROSS, J., PEACOCK, W. and EVANS, R. D. Metabolism of Radio-Iodine in the Thyroids of Rats Exposed to High or Low Temperatures. *Am. J. Physiol.*, **140,** 671, 1944.
   (c) DEMPSEY, E. W. and ASTWOOD, E. B. The Determination of the Rate of Thyroid Secretion at Various Environmental Temperatures. *Endocrinology*, **32,** 509, 1943.

10. MORIN, G. The Adrenal Medulla and Temperature Regulation. Reaction of Adrenal Secretion to Cold. *Rev. canad. biol.*, **5,** 388, 1946.

11. SCHAEFFER, G. Hormonal Factors Partaking in the Chemical Regulation of Temperature in Homeotherms. *Bull. Acad. Med.* (Paris), **130,** 587, 1946.

12. THÉRIEN, M. Contribution to the Physiology of Cold Acclimatization. *Laval Méd.*, **14,** Nos. 8 and 9 (Oct.–Nov.) 1949.

13. MANSFELD, G. and SCHEFF-PFEIFFER, I. Unknown Action of Thyroid in Regulation of the Body Temperature. *Arch. exp. Path. u. Pharmakol.*, **190,** 565, 1938.

14. MANSFELD, G. Hormonal Factors of Chemical Temperature Regulation and Two Hitherto Unknown Hormones of the Thyroid Gland. *Schweiz. med. Wchnschr.*, **72,** 1267, 1942.

15. SELYE, H. 'The Physiology and Pathology of Exposure to Stress.' Acta Inc., Montreal. 1950.

16. (a) DESMARAIS, A. and DUGAL, L. P. Peripheral Circulation and Adrenaline, Nor-Adrenaline Content of Suprarenals in White Rats Exposed to Cold. *Canad. J. Med. Sc.*, **29,** 90, 1951.

    (b) DESMARAIS, A. and DUGAL, L. P. Effect of Adrenaline and Nor-Adrenaline on Hypertrophy of Suprarenals in Rats Exposed to Cold. *Canad. J. M. Sc.*, **29,** 104, 1951.

    (c) DESMARAIS, A. and DUGAL, L. P. The Peripheral Circulation in the White Rat Exposed to Cold. *Rev. canad. biol.*, **9,** 206, 1950.

17. DUGAL, L. P. and THÉRIEN, M. The Influence of Ascorbic Acid on the Adrenal Weight during Exposure to Cold. *Endocrinology*, **44,** 420, 1949.

18. DUGAL, L. P. Effects of Cold, Ascorbic Acid and Age on 'formaldehyde-induced' Arthritis in the White Rat. *Canad. J. M. Sc.*, **29,** 35, 1951.

19. (a) THÉRIEN, M. and DUGAL, L. P. Urinary Excretion of Ascorbic Acid in the Rat and Guinea-Pig Exposed to Cold. *Rev. canad. biol.*, **8,** 248, 1949.

    (b) THÉRIEN, M. and DUGAL, L. P. Ascorbic Acid Content in the Tissues of the Rat partly Exposed to Intense Cold. *Rev. canad. biol.*, **8,** 440, 1949.

20. THÉRIEN, M., LE BLANC, J., HEROUX, O. and DUGAL, L. P. Effects of Ascorbic Acid on Several Biological Variables normally Affected by Cold. *Canad. J. Res.*, **27,** 349, 1949.

21. GIROUD, A., MARTINET, M. and BELLON, M. T. The Relationship between Ascorbic-Acid and Cortico-Adrenal Function in Man. Variation in the Excretion of Cortical Hormones. *C.R. Soc. Biol.*, **135,** 514, 1941.

22. SAYERS, G., SAYERS, M. A., LIANG, T. Y. and LONG, C. N. H. Cholesterol and Ascorbic Acid Content of Adrenal, Liver, Brain and Plasma following Haemorrhage. *Endocrinology*, **37,** 96, 1945.

23. HEROUX, O. and DUGAL, L. P. (a) Effect of Ascorbic Acid on Hypertension produced by Desoxycorticosterone Acetate .*Rev. canad. biol.*, **10,** 123, 1951.

    (b) Effect of Ascorbic Acid on Experimental Hypertension. *Canad. J. M. Sc.*, **29,** 164, 1951.

24. DUGAL, L. P. and FORTIER, G. Ascorbic Acid and Acclimatization to Cold in Monkeys. *J. Appl. Phys.*, **5,** 143, 1952.

25. SELLERS, E. A. and YOU, ROSEMARY. Prevention of Dietary Fatty Livers by Exposure to Cold Environment. *Science*, **110,** 713, 1949.

26. SELLERS, E. A. and YOU, S. S. Rôle of the Thyroid in Metabolic Responses to a Cold Environment. *Am. J. Physiol.*, **163**, 81, 1950.

27. SELLERS, E. A., YOU, S. S. and THOMAS, N. Acclimatization and Survival of Rats in the Cold: Effects of Clipping, of Adrenalectomy and of Thyroidectomy. *Am. J. Physiol.*, **165**, 481–485, 1951.

28. SELLERS, E. A., REICHMAN, S. and THOMAS, N. Acclimatization to Cold: Natural and Artificial. *Am. J. Physiol.*, **167**, 644, 1951.

29. SELLERS, E. A., REICHMAN, S., THOMAS, N. and YOU, S. S. Acclimatization to Cold in Rats: Metabolic Rates. *Am. J. Physiol.*, **167**, 651–655, 1951.

30. YOU, S. S., YOU, R. W. and SELLERS, E. A. Effect of Thyroidectomy, Adrenalectomy and Burning on the Urinary Nitrogen Excretion of' the Rat maintained in a Cold Environment. *Endocrinology*, **47**, 136–161, 1950.

31. YOU, R. W. and SELLERS, E. A. Increased O_2 Consumption and Succinoxidase Activity of Liver Tissue after Exposure of Rats to Cold. *Endocrinology*, **49**, 374–378, 1951.

32. BROWN, M. The Nutritional Status of the Eskimos. Fatty Metamorphosis in the Liver. *Proc. Can. Phys. Soc. Rev. canad. biol.*, **8**, 313, 1949.

33. ADOLPH, E. F. O_2 Consumption of Hypothermic Rats and Acclimatization to Cold. *Am. J. Physiol.*, **161**, 359–373, 1950.

34. ADOLPH, E. F. Committee on Geographical Exploration. Physiology Panel. Research and Development Board, April, 1948.

35. ADOLPH, E. F. Responses to Hypothermia in Several Species of Infant Mammals. *Am. J. Physiol.*, **166**, 75–91, 1951.

36. ADOLPH, E. F. Some Differences in Response to Low Temperature Between Warm Blooded and Cold Blooded Vertebrates. *Am. J. Physiol.*, **166**, 92–103, 1951.

37. HOLTKAMP, D. E., HILL, R. M., LONGWELL, B. B., RUTLEDGE, E. K. and BUCHANAN, A. R. Failure of Adrenal Cortical Hormones to Protect Against Cold in Young Normal and Adrenalectomized Rats. *Am. J. Physiol.*, **156**, 368–376, 1949.

38. COLLIP, J. B. and BILLINGSLEY, L. W. The Effect of Temperature upon Metabolism. *Tr. Am. Goiter A.*, 1936, p.121.

39. HART, J. S. Effects of Temperature and Work on Metabolism, Body Temperature and Insulation. Results with Mice. *Canad. J. Zool.*, **30**, 90–98, 1952.

40. KROGH, A., and KROGH, M. 'A Study of the Diet and Metabolism of Eskimos.' Copenhagen, 1913.

41. HEINBECKER, P. Studies on the Metabolism of Eskimos. *J. Biol. Chem.*, **80**, 461, 1928.

42. CRILE, G. W. and QUIRING, D. P. Indian and Eskimo Metabolisms. *J. Nutrition*, **18**, 361, 1939.

43. RABINOWITCH, I. M. and SMITH, F. C. Metabolic Studies of Eskimos in Canadian Eastern Arctic. *J. Nutrition*, **12**, 337, 1936.

44. HØYGAARD, A. 'Studies of the Nutrition and Physiological Pathology of the Eskimo.' Oslo, 1941.

45. LEVINE, V. E. Basal Metabolic Rate of Eskimo. *J. Biol. Chem.*, **133**, 61, 1940.

46. WILBER, C. G. and LEVINE, V. E. Fat Metabolism in Alaskan Eskimo. *Exper. Med. & Surg.*, **8**, 422, 1950.

47. BOLLERUD, J., EDWARDS, J. and BLAKELY, R. A. Basal Metabolism of Eskimos. Arctic Aeromedical Laboratory, 21-01-020, 1950.

48. BROWN, MALCOLM. Report on Queen's University, Kingston Ontario, Expedition, 1950.

49. HEINBECKER, P. Further Studies on the Metabolism of Eskimos. *J. Biol. Chem.*, **93**, 327, 1931.

50. GIAJA, J. and GELINEO, S. Nutrition and Resistance to Cold. *C.R. Acad. Sc.*, **198**, 2277, 1934.

51. DONHOFFER, S., and VONOTZKY, J. The Effect of Environmental Temerature on Food Selection. *Am. J. Physiol.*, **150**, 329, 1947.

52. STEFFANSON, V. (a) The Diets of Explorers. Mil. Surgeon, **95**, 1, 1944.
        (b) Pemmican. Mil. Surgeon, **95**, 89, 1944.

53. LOCKHART, E. E. Antarctic Trail Diet. *Proc. Am. Philos. Soc.*, **89**, 235, 1945.

54. FRAZIER, G. Acclimatization and the Effects of Cold on the Human Body as Observed at Little America III on the United States Antarctic Service Expedition 1939–41. *Proc. Amer. Philos. Soc.*, **89**, 249–255, 1945.

55. KEETON, R. W., LAMBERT, E. H., GLICKMAN, N., MITCHELL, H. H., LAST, J. H. and FAHNESTOCK, M. A. The Tolerance of Man to Cold as Affected by Dietary Modifications: Proteins v. Carbohydrates, and the Effect of Variable Protective Clothing. *Am. J. Physiol.*, **146**, 67–83, 1946.

56. MITCHELL, H. H., GLICKMAN, N., LAMBERT, E. H., KEETON, R. W. and FAHNESTOCK, M. A. The Tolerance of Man to Cold as Affected by Dietary Modifications: Carbohydrate v. Fat and the Effect of the Frequency of Meals. *Am. J. Physiol.*, **146**, 84–96, 1946.

57. SCHOENHEIMER, R. and RITTENBERG, D. Deuterium as an Indicator in the Study of Intermediary Metabolism. *J. Biol. Chem.*, **111**, 175,1935.

58. GLICKMAN, N., KEETON, R. W., MITCHELL, H. H. and FAHNESTOCK, M. A. Tolerance of Man to Cold as Affected by Dietary Modifications: High v. Low Intake of Certain Water Soluble Vitamins. *Am. J. Physiol.*, **146**, 538–558, 1946.

59. BLAIR, J. R., URBUSH, F. W. and REED, I. T. Observations on Diet and Nutrition. Armoured Medical Research Laboratory, Medical Department, Field Research Laboratory, 1947.

60. MITCHELL, H. H. and EDMAN, M. Nutrition and Resistance to Stress, with Particular Reference to Man. Office of the Quartermaster General, 1949.

61. AMES, A., GOLDTHWAIT, D. A., GRIFFITH, R. S. and MACHT, M. B. Methods of Rewarming. Quartermaster Climatic Research Laboratory, Environmental Protection Series. Report No. 134, 1948.

62. (a) HORVATH, S. M., FREEDMAN, A. and GOLDEN, H. Acclimatization to Extreme Cold. *Am. J. Physiol.*, **150**, 99, 1947.
    (b) HORVATH, S. M. and FREEDMAN, A. Effect of Cold on Efficiency. *J. Aviation Med.*, **18**, 158, 1948.
    (c) HORVATH, S. M. Ventilation of Clothing and Tolerance of Man to Low Environmental Temperatures. *J. Ind. Hyg. and Tox.*, **30**, 133, 1948.

63. (a) HORVATH, S. M., GOLDEN, H. and WAGER, J. Some Observations on Men Sitting Quietly in Extreme Cold. *J. Clin. Invest.*, **25**, 709, 1946.
    (b) HORVATH, S. M. and GOLDEN, H. Observations on Men Performing a Standard Amount of Work at Low Ambient Temperatures. *J. Clin. Invest.*, **26**, 311, 1947.

64. Adolph, E. F. and Molnar, G. W. Exchanges of Heat and Tolerances to Cold in Men Exposed to Outdoor Weather. *Am. J. Physiol.*, **146**, 507, 1946.

65. (a) Stein, H. J., Eliot, J. W. and Bader, R. A. Physiological Reactions to Cold and their Effects on the Retention of Acclimatization to Heat. *J. Appl. Physiol.*, **1**, 575, 1949.

(b) Stein, H. J., Bader, R. A., Eliot, J. W. and Bass, D. E. Hormonal Alterations in Men Exposed to Heat and Cold Stress. *J. Clin. Endocrinol.*, **9**, 529, 1949.

66. Bazett, H. C., Sunderman, F. W., Doupe, J. and Scott, J. Climatic Effects on the Volume and Composition of Blood in Man. *Am. J. Physiol.*, **129**, 69, 1940.

67. Burton, A. C., Scott, J. C., McGlone, B. and Bazett, H. C. Slow Adaptations in the Heat Exchanges of Man to Changed Climatic Conditions. *Am. J. Physiol.*, **129**, 84, 1940.

68. Scott, J. C., Bazett, H. C. and Mackie, G. C. Climatic Effects on Cardiac Output and the Circulation in Man. *Am. J. Physiol.*, **129**, 102, 1940.

69. Bazett, H. C. 'The Regulation of Body Temperatures in Physiology of Heat Regulation.' Newburgh, Saunders, Philadelphia, 1949.

70. (a) Glaser, E. M. Effect of Cooling and Heating on Warming of Skin and Body Temperature of Man. *J. Physiol.*, **109**, 366, 1949.

(b) Glaser, E. M. The Effects of Cooling and Warming on the Vital Capacity, Forearm and Hand Volume, and Skin Temperature of Man. *J. Physiol.*, **109**, 421, 1949.

(c) Glaser, E. M. Acclimatization to Heat and Cold. *J. Physiol.*, **110**, 330, 1950.

71. Carlson, L. D., Young, A. C., Burns, H. L. and Quinton, W. F. Acclimatization to Cold Environment. A.F. Technical Report No. 6247, March 1951.

72. Balke, B., Cremer, H. D., Kramer, K. and Reichel, H. Adaptation to Cold. *Klin. Wchnschr.*, **23**, 204, 1944.

73. Butson, A. R. C. Acclimatization to Cold in the Antarctic. *Nature*, **163**, 132, 1949.

74. Belding, H. S. Personal Communication, 1949.

75. Mackworth, N. H. Finger Numbness in Very Cold Winds. *J. Appl. Physiol.*, **5**, 533, 1953.

76. Brown, M. and Page, J. The Effect of Chronic Exposure to Cold on Temperature and Blood Flow of the Hand. *J. Appl. Phys.*, **5**, 221, 1952.

77. (a) Schmidt-Nielsen, K. Melting Point of Human Fats as Related to their Location in the Body. *Actaphysiol. Scandinav.*, **12**, 123–129, 1946.

(b) Schmidt-Nielsen, K. Determination of Melting Points in Human Fat. *Actaphysiol. Scandinav.*, **12**, 110–122, 1946.

O

CHAPTER 11

HYPOTHERMIA AND RESUSCITATION

In this section the changes that take place when body temperature falls will be considered in detail. The general effects of cold have to be distinguished from local effects or from local cooling.

As a preliminary, the term 'body temperature' needs to be defined and discussed. It is a useful concept to consider the body as consisting of a core, which is maintained at a relatively constant temperature, surrounded by an outer layer of tissue, the temperature of which may vary considerably according to the external environment, the degree of protection and the activity of the individual. There is a gradient of temperature from the inner core to the skin surface, the slope depending upon the above factors. In fact the concept of a deep core with a constant temperature is crude. There are considerable variations in the different parts of the deep core as Horvath (1) has clearly shown, e.g. the temperature of the liver might be 1° to 2°C (1·8° to 3·6°F) higher than the rectal temperature.

Rectal temperature cannot always be safely regarded as representing an average of deep body temperature. Bazett and his co-workers (2 a, b) have demonstrated that rectal temperature is influenced by the venous blood in the internal iliac veins. Such an effect is observed when legs are alternatively cooled and heated. Other sites have been proposed as more representative of body temperature, i.e. oral, intragastric or even intracardiac. The temperature of voided urine can also be used. Objections can be made against any of these suggestions as indeed for any one position. In a recent discussion at Copenhagen, it was agreed that the term 'body temperature' was misleading, when it was based on the measurement at any one locality. There are temperature differences always to be found, and these differences are not constant under varying conditions. It was agreed that it was better to measure the temperature at the selected site and to consider the phenomona recorded in relationship to the temperature of the site rather than confuse the issue by calling that temperature 'body temperature'. 'Measure the temperature in which you are interested'. Experiments concerned with the temperature of the hypothalamus should ideally include measurements of that region rather than measuring rectal temperature and considering that it will bear a constant relationship to hypothalamic temperature.

The Distribution of Temperature in the Body at Rest

Horvath (1) has measured intravascular temperatures at various sites in the dog. Amongst his interesting results, he has found that the temperature in the left auricle is higher than in the right auricle, i.e. there is a heat gain during the passage of blood through the pulmonary vessels. This heat gain is explained in part by the thermodynamics of gas exchange.

In man, no such complete mapping of temperature in different regions has been made so far. However, Bazett *et al.* (2 a, b) have measured the temperature of various intravascular sites as well as rectal temperature during exercise. Their results clearly show the marked difference between the temperature of blood from active muscles and rectal temperatures.

The technique of measurement of rectal temperature is also important. Mead and Bommarito (3) used a radiopaque catheter and took X-ray photographs of the position of the catheter tip. Owing to bending of the catheter, the tip lies posteriorly and to the right of the midline. As the catheter is pushed in for greater distances it usually bends and so the tip may not be at any greater

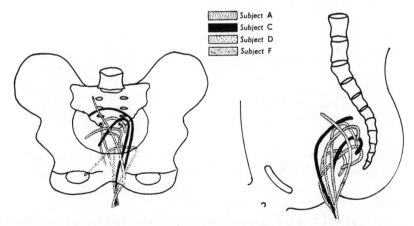

FIG. 65. The position of an 8-inch rectal catheter after seven separate insertions in four individuals. Note the variations in the position of the catheter tip. (From Mead and Bommarito, *J. Appl. Physiol.*, **2**, 97, 1949.)

distance from the surface (Fig. 65). It is implied in many papers that the farther a catheter is inserted the more reliable will be the temperature recorded. Mead and Bommarito have shown that this is not necessarily the case, as they measured temperature at various points along the catheter. Frequently higher temperatures were

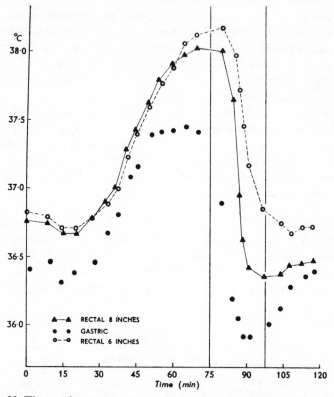

FIG. 66. The rectal temperature, recorded at two positions on a catheter inserted 8 inches, compared with gastric temperature. There is a marked difference between the readings taken 8 and 6 inches along the catheter, during body cooling. (From Mead and Bommarito, *J. Appl. Physiol.*, **2**, 97, 1949.)

recorded at parts other than the tip of the catheter. Where small temperature differences are relied upon as evidence of physiological change, it is clear that slight differences in the position of the catheter might be responsible (Fig. 66).

At present, summarizing the work on body temperature, it may be concluded that no one site or one reading of deep core temperature can give a figure which can be used to follow changes in the average temperature of the deep core. Theoretically the only reliable measure of body temperature is a calorimetric method following sudden death. Hart (4) has used such an approach in measuring heat content of small animals. In hypothermia he found in some cases rectal temperature lower than average body tem-

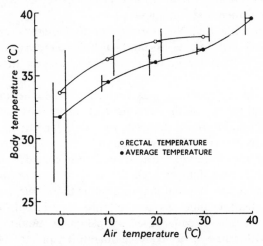

Fig. 67. Average body temperature (measured calorimetrically) and rectal temperature in mice killed at various environmental temperatures. The range of results is shown; at the lower temperatures there were instances where rectal was actually lower than average body temperature. (From Hart, *Science*, **131**, 325, 1951.)

perature, although in non-chilled animals killed at room temperature average body temperature was 1° to 2°C (2° to 3·6°F) below rectal (Fig. 67).

Hypothermia

Man and other animals can sustain a considerable fall in body temperature and survive. Hypothermia can be induced by exposure of the nude subjects to various environmental temperatures, by immersion in cold water, by packing the subject with ice bags, or by surrounding the subject with rubber tubing through which a refrigerant flows. All these methods have been employed, both experimentally and therapeutically.

One of the earliest recorded experiments is that of James Currie (5) over 150 years ago who measured mouth temperature in subjects whom he immersed in water at 7°C (44°F). He found an initial fall in mouth temperature followed by a slight rise associated with shivering. On removal from the bath and exposure to air, mouth temperature fell 1° to 1·5°C (2° to 3°F) and on subsequent immersion in a bath at 37°C (98°F) there was a further fall of 1° to 1·5°C (2° to 3°F). Mouth temperature rose when the subject was put into a bath at 43°C (109°F). Recent work has in general confirmed these findings.

It is probable that the duration of hypothermia affects the physiological adjustments, so the final body temperature reached is not the only factor of importance determining mortality or morbidity.

Interest in hypothermia was revived in 1935 because of its clinical application, in the treatment of neoplasms and also of schizophrenics. Smith (6) discusses the work done by himself and Fay using hypothermia (7). As tumours were said to occur chiefly in vascular and therefore warm areas, it was considered that low temperatures must damage such tissue more than normal tissue, especially as it was reported that embryonic tissue was damaged by relatively small changes in temperature. Local cold, 4° to 5°C (39° to 41°F), relieves pain caused by malignant tumours. Cooling patients with tumours to body temperatures of 32°C (90°F) or lower caused a fall in heart rate, blood pressure, and an abnormal T wave. Auricular fibrillation was occasionally noted. There was an initial haemo-concentration and leucocytosis with a slight fall in blood sugar. A curious complication was the occurrence of acute pancreatitis in a fairly large number of patients (8). Burton (9) observed some patients similarly treated and noted that re-warming was the dangerous phase with occasional circulatory failure. Dill and Forbes (10) who examined the effects of hypothermia found that relatively prolonged cooling, i.e. rectal temperature 24·5° to 32°C (76° to 90°F) for periods up to 48 hours caused a marked reduction in blood volume owing to plasma loss, with haemoconcentration. If re-warming was relatively rapid, then the intense peripheral vasoconstriction which compensated for the reduced blood volume in the chilled state was abolished with relative vasodilatation and circulatory failure. With slow re-warming there would be a shift of body fluid and the increase in blood volume could keep pace with the rise in body temperature.

Talbott (11, 12) and Dill and Forbes (10) examined the effects of hypothermia on blood chemistry and physiology in detail. Patients were cooled at the rate of 0·3° to 1·6°C (0·5° to 3·0°F) per hour to a rectal temperature of 26·5°C (80°F), maintained at approximately this level for periods up to 24 hours and rewarmed at the rate of 1°C (2°F) per hour. No serious after-affects were noted in survivors. There were two deaths in the series of 20 cases, which were due to cardiac failure during the hypothermic phase. During cooling there was an initial rise of heart rate and blood pressure, then both fell with rectal temperature. Venipuncture was very difficult owing to venous collapse; this is in contrast to the report on the Dachau experiments (q.v.), where it is stated that peripheral veins did not collapse and venipuncture was simple (13). Arterial constriction was also observed with prolonged hypothermia. The arm to leg circulation time was increased 2 to 3 times. Blood volume decreased with marked haemoconcentration, R.B.C. counts were increased by 25 per cent. Shivering was intense at the onset of cooling, and had to be controlled with sedatives. Occasional bouts of shivering were observed with rectal temperature as low as 24°C (75°F). Cardiac arrhythmias were common and auricular fibrillation was invariable when rectal temperatures of 26°C (80°F) were reached.

The horrible experiments at Dachau performed by the Nazis on prisoners in a concentration camp have been reported in detail by Alexander. As this report has had widespread publicity, it appears necessary to refer to a few of the details.

The prisoners were immersed in water at a temperature of 2° to 12°C (35° to 53°F). There was initial violent shivering, succeeded by intense muscular rigidity, which was only abolished at rectal temperatures below 27°C (81°F). Consciousness became clouded at rectal temperature of 31°C (88°F) and rectal temperatures reached 29·5°C (85·1°F) after 70 to 90 minutes of cooling: death occurred where rectal temperatures fell to 24° to 25·7°C (75° to 78·2°F). There appeared to be an inverse relationship between blood sugar and rectal temperature. As mentioned above, it was claimed that superficial veins did not collapse (Fig. 68).

There has been extensive animal work on the effects of hypothermia which throws light on the effects of cooling on man. Lutz (14) considered that anoxia was a major factor in causing death and claimed good results with oxygen. However, Noell (15) examined the EEG during hypothermia and concluded that the changes were not typical of anoxia or anaesthesia, but were similar to eserine poisoning. He suggested that the breakdown of acetyl choline might be delayed at low temperatures.

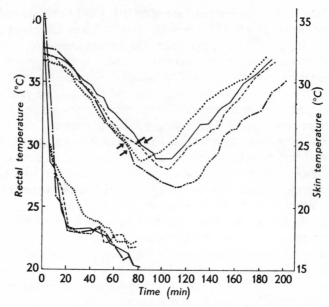

Fig. 68. Rectal temperature and skin temperature (ordinates) in Dachau prisoners immersed in water at 5°C (41°F). Time is in minutes. The arrows indicate the time of removal from the water. Note the continuing fall of rectal temperature. Compare with Fig. 77. (Redrawn from original graph in Alexander's Report.)

Crismon and Elliott (16) give the following figures for lethal body temperature in different species:

RECTAL TEMPERATURE

	Death	B.P. Failure
Marmot	0° to 5°C	
	32° to 41°F	
Rats	13° to 15°C	24° to 25°C
	55·4° to 59°F	75·2° to 77°F
Cats	14° to 16°C	22°C
	57·2° to 60·8°F	71·6°F
Dogs	18° to 20°C	22°C
	64·4° to 68°F	71·6°F
Man	24° to 26°C	
	75·5° to 78·8°C	

These workers considered that the cardiac failure, which appears to be the main cause of death in hypothermia, was due to the direct effect of cold on the pacemaker of the heart. They argued that local heating of the heart, i.e. of the pacemaker should avert or delay

cardiac failure, so they inserted a small electric heater in the oeso-
phagus of a rat in close proximity to the heart. The rat was cooled:
when the blood pressure began to fall and the heart rate was very
slow, the heater was turned on. The blood pressure rose and heart
rate increased: there was also a check in the rate of fall of body
temperature, although the quantity of heat supplied was very

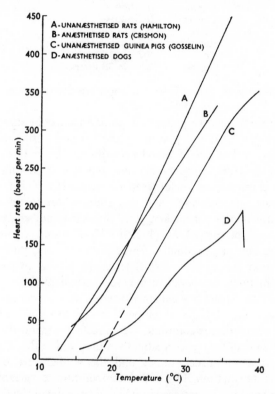

FIG. 69. The relationship between body temperature and heart rate in different
species. (From Hegnauer and Penrod, *A.E. Tech. Report*, 5912, 1950.)

small. Crismon and Elliott then tried a similar technique in dogs,
but the results were not encouraging.

In the rat there is a linear relationship between rectal temperature and heart rate. The blood pressure rises slightly in the
initial phase of cooling, then at rectal temperatures of 30° to 23°C
(86° to 73°F) it declines slowly followed thereafter by a rapid fall
(17, 18) (Fig. 69). Crismon and Elliott (19) found that the B.P. in

the hypothermic rat was maintained better after the administration of lanatoside C.

There is a marked species variation in the response to hypothermia as shown by Crismon and Elliott. Horvath *et al.* (20) expressed this by measuring survival time at a constant air temperature of −35°C (−31°F), i.e.:

Mouse	0·4 hours
Canary	0·6 hours
Rat	0·75 to 2·0 hours
Rabbit	3·5 to 6·5 hours
Chicken	3·3 to 16 hours
Pigeon	22 to 78 hours

Some aspects of this species variation are discussed in Chapter 10.

The cardiovascular response to hypothermia may be summarized as follows (21, 22): there is an initial marked peripheral vasoconstriction with a rise in blood pressure and heart rate. During this stage the rectal temperature does not fall. If cooling is continued or is intensified, rectal temperature starts to fall. As it does so the heart rate decreases, and this effect is due to the lowered temperature of the pacemaker. The slowing is not abolished by atropine or vagotomy (23, 24) (Fig. 70). Blood pressure falls during this period but only gradually. There is an accompanying haemoconcentration due to a fluid shift from the plasma to the tissue. The fall in body temperature becomes exponential and follows Newton's Law. Heart rate becomes progressively slower and arrhythmias appear, and at rectal temperatures of 30°C (86°F) or lower auricular fibrillation is common. When rectal temperature reaches a level of 25°C (77°F) or thereabouts in man and in the dog, blood pressure may fall precipitously and the animal dies owing to ventricular fibrillation. This lethal temperature is variable, as is demonstrated by a recent case in which a woman survived a fall in rectal temperature to 18°C (64·4°F) or possibly even lower (25).

Metabolism and Respiration

During the initial phase of cooling, there is intense shivering except in anaesthetized animals. It is worth emphasizing the complication introduced by anaesthesia: shivering is suppressed, and the anaesthetic effect may be prolonged as Fuhrman (26) has shown.

In the unanaesthetized animal during shivering there is a marked

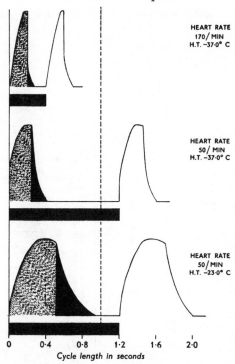

FIG. 70. Left ventricular pressure tracings in a normal dog (top tracing) and a hypothermic dog (bottom tracing). H.T. = heart temperature. The middle tracing was obtained from a dog whose heart rate had been slowed by vagal stimulation. The marked prolongation of both systole and relaxation in hypothermia is evident. (From Hegnauer *et al.*, *Am. J. Physiol.*, **161**, 455, 1950.)

rise in oxygen consumption and occasionally a slight rise in rectal temperature. In man this effect can be very marked. Behnke exposed himself and two other volunteers to water temperatures of 5·5° to 10°C (42° to 50°F) (27). In water at 6·1°C (43°F) rectal temperature first rose slightly and then fell in a linear fashion to 35·8°C (96·5°F) at the end of an hour. Metabolic rate increased from 85 kg.cal/hr to approximately 500 kg.cal/hr. This increase was relatively slow, taking some 30 minutes to reach a maximum. In spite of this great increase in metabolic rate, heat loss was some 30 to 36 per cent greater, as computed by Burton's formula (Fig. 71). The intensity of shivering was related to rectal temperature, and violent shivering did not start until rectal temperature began to fall.

It is interesting to compare these results with Molnar's (28)

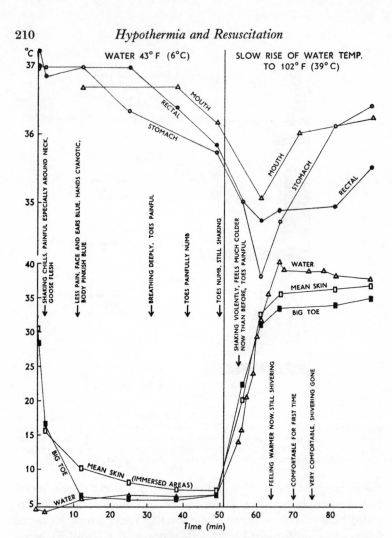

Fig. 71. Changes in body and skin temperatures of subject immersed in water at 6°C (43°F) for 52 minutes. The water was then warmed to 39°C. Note the sharp fall of gastric oral and rectal temperatures initially on warming. (From Behnke and Yaglou, *J. Appl. Physiol.*, 3, 59, 1950.)

calculations for survival at sea. He estimates that immersion for 1 hour at 4·4°C (40°F) will kill 50 per cent of men immersed. At 6·1°C (43°F) 50 per cent survival time would be approximately 75 minutes. It is clear that Behnke and his subjects would have survived a much longer period. This calculation of Molnar's was based in part on the Dachau experiments, the subjects of which were prisoners and unlikely to have been well nourished (Fig. 72).

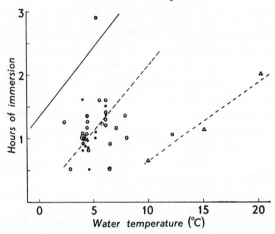

FIG. 72. Survival times for immersion in cold water for different times. The heavy continuous line indicates the time after which few survivors could be expected; the thin dashed line, all would survive; and the thick dashed line indicates 50% survival. Data derived from Dachau experiments (● fatal results, O survival) and from Spealman's data, ▷. (From Molnar, *J.A.M.A.*, **131**, 1046, 1946.)

Spealman (29) followed the cooling rate in animals and man at different water bath temperatures. There were differences noted between the human subjects, which was also observed by Pugh *et al.* (Fig. 73).

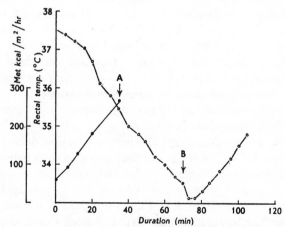

FIG. 73. Rectal temperature and oxygen consumption of a normal subject immersed in water at 15°C. The arrow (B) indicates removal from bath. In spite of the steep rise in metabolism, the rectal temperature falls steadily. (From Spealman, *Am. J. Physiol.*, **146**, 262, 1946.)

Pugh *et al.* (30) have recently re-examined the rate of cooling in water, and have demonstrated a very considerable individual variation, which appears to be related to the thickness of the subcutaneous fat. In one subject, with a very thin layer of fat (average approximately 5 mm. over the trunk) the cooling rates were very similar to those reported at Dachau. In another subject, who was an expert long-distance swimmer, cooling rates were extremely slow and there was substantially no fall in rectal temperature in water at 15°C (59°F). He had a layer of fat up to 3 cm. thick over his trunk. A third subject, who appeared to be lean, also cooled slowly; his fat layer proved to be of the order of 1·5 cm.

The expert swimmer swam in water at 15°C (59°F) for 6 hours with no fall of rectal temperature. On Molnar's chart, the 50 per cent survival time would have been only 3 hours 15 minutes, and few, if any, survivors would have been expected after 5·5 hours. On the other hand, 22 people recently swam the English Channel in times varying from 9 to 15 hours at water temperatures of the order of 15·5° to 16·5°C (60° to 62°F). There was no evidence of severe hypothermia in these swimmers, all of whom appeared to be fat. It is customary for long-distance swimmers to cover their bodies with a layer of grease (lanolin) and it is claimed that this delays body cooling. The quantity of lanolin normally employed would, however, only provide a uniform layer over the body of approximately 1 mm. thick; and after a short period of swimming much of this is washed off. It is difficult to understand how the extra insulation provided by such a thin layer of fat can play a significant rôle in reducing heat loss. However, interviews with long-distance swimmers show that they are convinced of the importance of fat, and it is possible that other factors are involved. This is a problem that deserves further investigation.

Glaser (31) has emphasized that it is important to swim rather than to float motionless in cold water, in order to survive after shipwreck: he considers that the heat production during swimming would be high enough to balance or almost to balance heat loss. However measurement of the metabolic cost of sustained swimming gives figures of the order of 300 to 350 kcal/m²/hr, which is not greatly in excess of the heat production from shivering, at any rate in peak periods. Heat loss from a moving body as compared with a subject keeping still would possibly be increased, and for thin subjects keeping still might be better than swimming. The evidence on this point is not yet adequate.

All observations on hypothermia suggest that the increase in metabolic rate with cooling does not persist after rectal temperature has fallen to approximately 35°C (95°F). Thereafter there is a gradual decline to reach basal values or lower at rectal temperatures of 27° to 30°C (80·6° to 86°F). Shivering usually ceases at rectal temperatures of 30° to 33°C (86° to 91·4°F) and is succeeded by a persistent muscular rigidity. The same sequence has been observed in dogs by Hegnauer and Penrod (24, 32, 33, 34). During re-warming, shivering begins again on reaching rectal temperatures above 30°C (86°F).

Respiration is increased at the onset of hypothermia and there is a fall in alveolar CO_2. Subsequently the minute volume falls, although oxygen consumption increases. Grosse-Brockhoff (35) showed that the excitability of the respiratory centre is increased at body temperatures of 34·5°C (94°F) and then gradually decreases. When body temperature has fallen to 26°C (78·8°F) in the dog, carbon dioxide no longer acts as a respiratory stimulant, but depresses the centre. Oxygen deficiency can still increase respiration at this stage.

The CO_2 content of both arterial and venous blood drops in hypertension according to the Dachau experiments (13). Hegnauer and Penrod (24) followed oxygen and CO_2 content of both arterial and venous blood in their experiments on hypothermia in dogs. At body temperatures of 30°C (86°F), arterial oxygen content was increased by approximately 25 per cent, due largely to an increased haematocrit. There was also a fall of some 15 per cent in arterial CO_2 content. On cooling the animals further the oxygen content was either equal to initial values before hypothermia; in spite of the great fall in respiration venous oxygen changes approximately paralleled arterial O_2. The carbon dioxide content fell but not dramatically: certainly the changes were less than those reported at Dachau. However, in those animals which survived to rectal temperatures of 20°C (68°F), the oxygen content of the arterial blood fell and the CO_2 rose. This is the stage of respiratory failure.

In dogs breathing O_2 (36, 37), the CO_2 rose steeply at rectal temperatures of 25° to 20°C (77° to 68°F). It is pointed out by the authors that breathing oxygen, the volume of dissolved oxygen at 22°C (71·6°F) would be 2·7 ml. O_2/100 ml. blood. The oxygen consumption is less than 2·7 ml./kg/min., so until cardiac output fell below 100 ml./min./kg the entire metabolic demands of the animal could be met by the dissolved oxygen, without any dissociation of haemoglobin. This would mean that the CO_2 would increase. Animals breathing oxygen died at materially higher temperatures than those breathing air. CO_2 is depressant at low body temperatures (35) and this may account for the early respiratory failure of dogs breathing oxygen. Bigelow *et al.* (38) have also concluded that oxygen inhalation is of no value, so the weight of the evidence is against the use of oxygen therapy in hypothermia.

Many workers have pointed out that the dissociation curve of haemoglobin will be shifted to the left with a fall in temperature, and on theoretical grounds Werz (39), for example, concluded that very low partial pressures of oxygen would be found in the tissues in hypothermia. Hegnauer and Penrod (24) point out that a shift in the pH of blood to the acid side could counteract the effect of temperature on the dissociation of haemoglobin. Their results show that a progressive fall in pH does occur during hypothermia, and is of sufficient magnitude significantly to counteract the effect of temperature, so the dissociation curve is not greatly shifted to the left, and haemoglobin still releases oxygen at comparatively high partial pressures.

Other Metabolic Changes in Hypothermia

The blood sugar level varies considerably during the different stages of hypothermia. In the Dachau experiments it was found that the blood sugar varied inversely with body temperature. As rectal temperature fell to 28°C (82°F) or lower, the blood sugar rose to a peak approximately double the control level, and during re-warming the level fell again.

Many years ago, Finney, Dworkin and Cassidy (40) showed that shivering in dogs immersed in ice water could be inhibited by lowering the blood sugar with insulin. Talbott (12) confirmed this, showing that in man, hypoglycaemia abolished shivering. According to Fuhrman and Crismon (41), hyperglycaemia during body cooling was reported by Claude Bernard. The blood sugar response to cooling in rats appears to depend on the rate of cooling (38). In acute hypothermia liver glycogen is rapidly depleted and there is hyperglycaemia, but with gradual cooling hypoglycaemia is the rule. Elliott and Crismon (42) reported that starved rats with low initial liver glycogen survived to lower body temperatures than control animals. With longer initial starvation, liver glycogen is raised, and such animals do not survive cooling. If glucose is given during hypothermia survival time is increased. These findings are explained by Elliott and Crismon (42) on the basis of the effects of potassium. When glycogen is converted to glucose in the liver, potassium is also produced. There is an increased sensitivity to potassium in the cooled animal: giving glucose or calcium will protect cooled rats against fatal levels of potassium.

Grosse-Brockhoff (35) reported a slight rise in blood sugar content in hypothermia during the initial phase of excitation, but with the decline in metabolic rate with further cooling there is a fall in blood sugar. Deuticke (43) showed that at temperatures of 27° to 25°C (80·6° to 77°F) phosphorylysis in striated muscle is impaired.

The blood flow through the liver, as judged by the removal of bromsulphalein, is reduced in hypothermia (Hegnauer and Penrod (24)). Detailed studies are required on the changes in enzyme systems during hypothermia.

The effect of hypothermia on the endocrine glands has not been followed in great detail. Hegnauer and Penrod (24) have examined the part played by the thyroid. Dogs fed on 4 mg/kg/thyroxin for 13 days cooled faster than on previous occasions. In rats rendered hypothyroid the cooling rate was similar to that of the control rate.

THE EFFECT OF THYROID EXTRACT ON THE COOLING RATE OF THE DOG

	No. Dogs	Mean Cooling Rate to 20°C °C/min.
First immersion	6	0·173 ± 0·0555
Second immersion	5	0·150 ± 0·0478
After thyroid therapy	4	0·243 ± 0·0269

Resuscitation

The proper treatment of hypothermia can be summed up as either slow or rapid re-warming, never moderate re-warming.

Rapid re-warming was first advocated by Laptschinski, in 1880: but the credit for the re-discovery of this mode of treatment must go to the German workers who first demonstrated the effects clearly on animals. Weltz and his co-workers (44) chilled guinea-pigs to body temperatures of 18°C (64·4°F) and lower, and then dipped the animals into water at 45° to 50°C (113° to 122°F). Body temperature rose rapidly and the animals recovered. Experiments were then carried out on larger animals, including pigs.

P

Finally, the method was tested on human victims by Rascher in the Dachau experiments (13).

It was frequently reported during the war that survivors rescued from the sea and who were alive at the time of rescue, died shortly after they were taken out of the water. Treatment usually consisted of wrapping in warm blankets, hot drinks, and if available an electric cradle. When similar treatment is given to chilled animals it is found that the initially low rectal temperature falls even further during the re-warming phase, and death may ensue.

The fall of rectal temperature on re-warming was first observed by James Currie in 1790 (5) and has frequently been reported since that time. The skin and superficial tissues are, at the time of removal from the cold water, much colder than the deep tissues. The skin temperature will be approximately the same as the water temperature, so the gradient from the core of the body to the skin may be of the order of 20°C (36°F). The blood vessels in the superficial parts of the periphery are intensely constricted and the blood flow greatly reduced. On re-warming, skin temperature rises rapidly and the blood flow through skin and muscle increases. There will be a steep fall in temperature of the blood flowing through this chilled zone and hence the temperature of the venous blood returning to the heart is reduced. The gastric temperature may fall 3°C (5·4°F) in 5 minutes (Fig. 71). Such a rapid drop in cardiac temperature may produce cardiac arrest or ventricular fibrillation. Death in most cases appears to be due to cardiac failure. However, when chilled animals or man are placed in water at a temperature of 45° to 50°C (113° to 122°F), the periphery is re-warmed very rapidly, hence the cooling effect on the peripheral circulating blood is transient. The fall in rectal or gastric temperature is still observed, but it lasts only a few minutes and may be less than 1°C (1·8°F).

The German work has been confirmed and extended by Hegnauer and Penrod (24), Behnke and Yaglou (27), Bigelow *et al.* (38) on animals and on man.

The work so far described shows that non-hibernating homeotherms cannot survive at rectal temperatures much below 15°C (59°F). Certainly rats cooled below 12°C (53·6°F) invariably die and no resuscitation procedure has proved successful.

Recently, however, some remarkable experiments have been reported by Andjus (45). He has succeeded in cooling rats to a rectal temperature of 1°C (33·8°F), maintaining them at this level for up to 1 hour, and has then resuscitated the animals. During the period when this very low temperature is maintained there is no evidence of any cardiac or respiratory activity, the electrocardiogram being flat. As these experiments are obviously of the greatest importance, a full description will be given. It may be added that one of us (O.G.E.) has witnessed these experiments, and has been given a complete account of the procedure, which is extremely simple to repeat.

The unanaesthetized rat is placed in a two-litre flask which is corked and put in a refrigerator kept at approximately 5°C (41°F). The animal is therefore exposed to cold plus a gradual increasing concentration of CO_2. After 1 to 2 hours, the rectal temperature will have dropped to approximately 18°C (64·4°F). The rat is removed from the refrigerator and placed in a dish filled with ice, which is packed all around the animal. The rectal temperature falls rapidly and within some 10 minutes reaches 1°C (33·8°F). The animal is left in the ice for about an hour and it is then removed. Heat is applied to the chest wall over the heart by means of a metal spatula heated in a flame. The spatula is applied at frequent intervals, and artificial respiration given in between. After some 10 to 15 minutes of this treatment, occasional heart-beats can be detected. At this stage the rectal temperature has usually risen to approximately 10°C (50°F) owing to direct re-warming of the tissues by the environmental temperature of the laboratory (20°C (68°F)). As soon as heart-beats are definitely established, the neck is warmed with water at 40°C (104°F) intermittently. Spasmodic gasping movements are usually observed at this stage, and the whole animal is then re-warmed in water at 40°C (104°F). Respiratory movements gradually become more regular as the body temperature rises, and when this reaches approximately 27° to 28°C (80·5° to 82·5°F), the animal is removed from the water bath, dried and put in a hot air oven, kept at approximately 37°C (98·6°F). As the body temperature rises, consciousness is regained and the rat starts to make voluntary movements. The time elapsing between removal from the ice and voluntary activity is approximately 1 hour. For some 24 to 28 hours after resuscitation, temperature regulation is impaired, and the rat is ataxic. Thereafter recovery appears to be complete.

These remarkable experiments open up a new field for research on the effects of hypothermia. There is no doubt that the technique employed by Andjus can be improved and it will be necessary to repeat a similar procedure in other animals. The biochemical changes in these animals should prove of the greatest interest, as well as the hormonal adjustments.

In acute hypothermia, the treatment of choice consists of immediate rapid re-warming in water at a temperature of 45°C (113°F). The problem of chronic hypothermia is less straightforward. Acute hypothermia may be defined as the rapid cooling that occurs with immersion in cold water, with body temperature falling steadily. Survivors of acute hypothermia are unlikely to have been chilled for more than 12 hours.

Chronic hypothermia may be the result of exposure to a cold environment other than water; it has been observed carefully in therapeutic cooling (10, 11), when patients were kept for many hours at a low body temperature. Otherwise chronic hypothermia may occur in cold climates in which people have been exposed inadequately protected. Even the most effective Arctic clothing so far designed will not prevent loss of stored heat at low temperatures if the subject is at rest. Hence serious chronic hypothermia may be observed on land during military campaigns, in therapeutic work, or occasionally as accidents to civilians.

Body temperature falls more slowly and total exposure to cold may greatly exceed 12 hours. Probably the most important difference between acute and chronic hypothermia is in the change of blood volume. The circulating blood volume decreases, owing to plasma loss, and there is considerable haemoconcentration. According to Talbott (11), the fall in blood volume is progressive, so there is no sharp distinction between acute and chronic hypothermia in this respect: the difference is to a certain extent arbitrary. There is intense peripheral constriction: peripheral resistance may be increased 4 or 5 times (46). When re-warming begins, the peripheral vessels dilate and the peripheral resistance falls. Cold blood returns to the heart, which is slowed still further, with a reduction in cardiac output. As both resistance and output fall, there is a dangerous drop in blood pressure.

There is still some controversy as to the mode of treatment in cases of chronic hypothermia with a greatly reduced blood volume. Grosse-Brockhoff and others consider that rapid re-warming is again the treatment of choice. Others, including Talbott and Burton (47), believe that circulatory collapse complicates rapid re-warming. In therapeutic hypothermia when very low rectal temperatures of 25° to 27°C (77° to 80·6°F) were maintained for as long as 48 hours, re-warming was carried out with great caution. Refrigeration was stopped and patients re-warmed at room temperatures of 21° to 24°C (70° to 75°F) at the rate of approximately 0·55°C (1°F) per hour. No extra heat was supplied to the patient, who re-warmed by his own metabolic efforts. It was considered that blood volume was gradually restored during the slow re-warming, although detailed evidence is lacking on this point.

There is another important difference between acute and chronic hypothermia emphasized by Grosse-Brockhoff (35). In acute hypothermia the period of increased metabolism with violent shivering is comparatively short owing to the rapid drop in rectal temperature. In chronic hypothermia, with a gradual fall in temperature, shivering persists for a long time with a sustained rise in cardiac output and blood pressure. There is therefore exhaustion of liver and muscle glycogen and also a fall in glycogen content of cardiac muscle. So cardiac failure may result at comparatively high temperatures owing to the reduction of cardiac glycogen, and cardiac injuries from hypothermia may persist. There is also evidence of damage to the adrenals in prolonged hypothermia. This point is considered in more detail in the discussion on the rôle of ascorbic acid (see Chapter 10). Owing to the depletion of glycogen, intravenous glucose should be given during resuscitation. Many drugs have been tried in the treatment of hypothermia, but none appears useful. In chronic hypothermia, owing to the cardiac damage, strophanthin or digitalis may possibly prove of value.

In summary, in acute hypothermia, the main danger is a further fall in body temperature; rapid re-warming is the most effective treatment. In chronic hypothermia body temperature does not fall so fast, but the duration of exposure produces marked changes in blood volume and depletion of glycogen reserves. Rapid re-warming with intravenous glucose and specific cardiac therapy is one suggested method of treatment. The alternative is a very slow rise in body temperature, possibly combined with glucose. Moderate re-warming is definitely to be condemned.

The case of a negress who survived cooling to a rectal temperature of 18°C (64·4°F) has recently been described in detail by Laufman (25). This woman, who was intoxicated, lay unconscious for 11 hours out of doors at an air temperature of −18° to −24°C (−0·4° to −11·2°F). One-and-a-half hours after admission to hospital her rectal temperature was 18°C (64·4°F). No active re-warming was undertaken, and she lay naked at a room temperature of 20°C (68°F). Her rectal temperature rose steadily at the rate of 1°C (1·8°F) per hour. She suffered from severe frostbite, transient hypertension and renal damage, but made a good recovery, There are many striking features of this case, including survival from a rectal temperature much lower than any previously recorded, and the successful treatment of gross hypothermia by slow re-warming without any heating. The argument used above concerning slow versus rapid re-warming stressed the diminished blood volume with haemoconcentration and the danger of circulatory collapse if re-warming were rapid. Although blood volume studies were not done in this patient, her haematocrit shortly after admission was only 33, with 3½ million red blood cells and 10·9 gr. haemoglobin. There certainly was no evidence of haemoconcentration, which casts doubt on any substantial diminution in blood volume. The blood pH was 7·17, which supports the findings of Hegnauer and Penrod (24) concerning an acidaemia in hypothermia, so assisting the dissociation of haemoglobin. The blood sugar was 438 mgs. per cent which fits in with the observations cited above regarding hyperglycaemia in hypothermia. The full account of this case should be studied for further details.

REFERENCES

1. HORVATH, S. M., RUBIN, A. and FOLTZ, E. L. Thermal Gradients in the Vascular System. *Am. J. Physiol.*, **161**, 316, 1950.
2. (a) BAZETT, H. C., MENDELSON, E. S., LOVE, L. and LIBET, B. Precooling of Blood in the Arteries, Effective Heat Capacity and Evaporative Cooling as Factors Modifying Cooling of the Extremities. *J. Appl. Physiol.*, **1**, 169, 1948.
   (b) BAZETT, H. C., LOVE, L., NEWTON, M., EISENBERG, L., DAY, R. and FORSTER, R. Temperature Changes in Blood Flowing in Arteries and Veins in Man. *J. Appl. Physiol.*, **1**, 3, 1948.
3. MEAD, J. and BOMMARITO, C.L. Reliability of Rectal Temperature as an Index of Internal Body Temperature. *J. Appl. Physiol.*, **2**, 97–109, 1949.
4. HART, J. S. Average Body Temperature in Mice. *Science*, **113**, 325, 1951.
5. CURRIE, J. The Effects of Water, Cold and Warm, as a Remedy in Fever. Appendix on the Treatment of Shipwrecked Mariners. Liverpool, 1798.
6. SMITH, G. W. Use of Cold in Medicine. *Ann. Int. Med.*, **17**, 618, 1942.
7. FAY, T. and SMITH, G. W. Observations on Reflex Response During Prolonged Periods of Human Refrigeration. *Arch. Neurol. and Psychiat.*, **45**, 215–222, 1941.
8. VAUGHAN, A. M. Experimental Hibernation and Metastatic Growths. *J.A.M.A.*, **114**, 2293, 1940.

9. BURTON, A. C. Unpublished Observations, 1942.
10. DILL, D. B. and FORBES, W. H. Respiratory and Metabolic Effects of Hypothermia. *Am. J. Physiol.*, **132**, 685, 1941.
11. TALBOTT, J. H. Physiology of Hypothermia. *New England J. Med.*, **224**, 281, 1941.
12. TALBOTT, J. H., CONSOLAZIO, W. V. and PECORA, L. J. Hypothermia. *Arch. Int. Med.*, **68**, 1120, 1941.
13. ALEXANDER, L. (*a*) Treatment of Shock from Prolongd Exposure to Cold Water. Publication Board Report No. 250, 1946. Department of Commerce, Washington, D.C.
    (*b*) Medical Science Under Dictatorship. *New Eng. J. Med.*, **241**, 39, 1949.
14. LUTZ, W. The Experimental Adaptation of Warm-Blooded Animals to a Cold-Blooded Existence. A Contribution to the Mechanism of Death by Cold. *Klin. Wchnschr.*, **22**, 727–733, 1943.
15. NOELL, W. Hypothermia and Central Nervous System. *Arch. Psychiat.*, **118**, 29, 1945.
16. CRISMON, J. M. and ELLIOTT, W. H. Circulatory and Respiratory Failure in the Hypothermic Rat. *Stanford M. Bull.*, **5**, 115, 1947.
17. ADOLPH, E. F. Oxygen Consumption of Hypothermic Rats. *Am. J. Physiol.*, **161**, 359, 1950.
18. CRISMON, J. M. Hypothermia on the Heart Rate, the Arterial Pressure and the Electrocardiagram of the Rat. *Arch. Int. Med.*, **74**, 253, 1944.
19. CRISMON, J. M. and ELLIOTT, W. H. Effect of Lanatoside C upon the Survival of Rats Subjected to Severe Hypothermia. *Am. J. Physiol.*, **151**, 221–228, 1947.
20. HORVATH, S. M., FOLK, G. E., CRAIG, F. N. and FLEISHMANN, W. Survival Time in Cold. *Science*, **107**, 171, 1948.
21. KÖNIG, F. H. Physiological Observations in Hypothermia. *Klin. Wchnschr.*, **22**, 45, 1943.
22. KÖNIG, F. H. Blood Temperature and Heat Regulation. *Pflügers Arch.*, **247**, 497, 1944.
23. HATERIUS, H. O. and MAISON, G. L. Hypothermia in Dogs, Lightly Anaesthetized. U.S.A.F., MCREXD. 696–113, 5th Feb. 1948.
24. HEGNAUER, A. H. and PENROD, K. E. The Hypothermic Dog. A.F. Tech. Report, 5912, Feb. 1950.
25. LAUFMAN, H. Profound Accidental Hypothermia. *J.A.M.A.*, **147**, 1201, 1951.
26. FUHRMAN, F. A. Effect of Body Temperature on Drug Action. *Physiol. Rev.*, **26**, 247, 1946.
27. BEHNKE, A. R., and YAGLOU, C. P. Response of Man to Chilling and Re-Warming. *J. Appl. Physiol.*, **3**, 591, 1950.
28. MOLNAR, G. W. Survival of Hypothermia by Men Immersed in the Ocean. *J.A.M.A.*, **131**, 1046–1050, 1946.
29. SPEALMAN, C. R. Body Cooling of Rats, Rabbits and Dogs following Immersion in Water, with a few Observations on Man. *Am. J. Physiol.*, **146**, 262–266, 1946.
30. PUGH, L. G. C., MOTTRAM, R. and EDHOLM, O. G. Unpublished Observations, 1951.
31. GLASER, E. M. Immersion and Survival in Cold Water: Heat Production during Swimming. *Nature*, **166**, 1068, 1950.
32. HEGNAUER, A. H., D'AMATO, H. and FLYNN, J. The Influence of

Intra-Ventricular Catheters on the Course of Immersion Hypothermia in the Dog. *Am. J. Physiol.*, **167**, 1, 63, 1951.

33. WOLFF, R. C. and PENROD, K. E. Factors affecting Rate of Cooling in Immersion Hypothermia in the Dog. *Am. J. Physiol.*, **163**, 3, 580, 1950.

34. HEGNAUER, A. H., FLYNN, J. and D'AMATO, H. Cardiac Physiology in Dog during Re-Warming from Deep Hypothermia. *Am. J. Physiol.*, **167**, 1, 69, 1951.

35. GROSSE-BROCKHOFF, F. Pathologic Physiology and Therapy of Hypothermia. Chapter VIII E., 'German Aviation Medicine, World War II', Surg. Gen., U.S.A.F., **2**, 828–842, 1950.

36. PENROD, K. E. and FLYNN, J. Cardiac Oxygenation during Severe Hypothermia in the Dog. *Am. J. Physiol.*, **164**, 1, 79, 1951.

37. ROSENHAIN, F. R. and PENROD, K. E. Blood Gas Studies in the Hypothermic Dog. *Am. J. Physiol.*, **166**, 1, 55, 1951.

38. BIGELOW, W. G., LINDSAY, W. F. and GREENWOOD, W. F. Hypothermia, its possible Rôle in Cardiac Surgery: Investigation of Factors governing Survival in Dogs at Low Temperatures. *Ann. Surg.*, **132**, 849, 1950.

39. WERZ, R. Anoxia as a Cause of Death from Cold. *Arch. exp. Path. u. Pharmakol.* Leipzig, **202**, 561–593, 1943.

40. FINNEY, W. H., DWORKIN, S. and CASSIDY, G. J. Low Body Temperature and Insulin on Respiratory Quotients. *Am. J. Physiol.*, **80**, 301–310, 1927.

41. FUHRMAN, F. A. and CRISMON, J. M. The Influence of Acute Hypothermia on the Rate of Oxygen Consumption and Glycogen Content of the Liver and on Blood Glucose. *Am. J. Physiol.*, **149**, 552–559, 1947.

42. ELLIOTT, W. H. and CRISMON, J. M. Increased Sensitivity of Hypothermic Rats to Injected Potassium and the Influence of Calcium Digitalis and Glucose on Survival. *Am. J. Physiol.*, **151**, 366–372, 1947.

43. DEUTICKE, H. J., Sedimentation Constants of Muscle Proteins. Hoppe-Seyler, **224**, 216, 1934.

44. WELTZ, G. A., WENDT, H. J. and RUPPIN, H. Warming after Severe Hypothermia. *Munch. Med. Wchnschr.*, **89**, 1092, 1942.

45. ANDJUS, R. The Resuscitation of Adult Rats Cooled Close to Freezing. *C.R. Acad. Sc.*, **232**, 1591, 1951.

46. THAUER, R. and WEZLER, K. Metabolism as a Function of Temperature Regulation. *Ztschr. ges. exper. Méd.*, **112**, 95, 1948.

47. Cold Injury. Transactions of First Conference. Josiah Macy Foundation, New York, 1952.

CHAPTER 12

LOCAL COLD INJURY

In a previous chapter the effects of general body cooling were considered. The commonest form of cold injury is essentially a local effect, and includes frostbite, trench-foot or immersion foot, and, mildest of all, the chilblain. The first question is 'Are low temperatures greatly injurious to tissues or individual cells?' If so, 'what ranges of temperatures are dangerous?' And finally, 'why and how are low temperatures injurious?'

When the effect of changes of temperature on isolated cells or tissues are considered, it is clear that irreversible physico-chemical changes take place when temperatures are raised. Coagulation of protein at temperatures of 50° to 60°C (122° to 140°F) will kill the cell. So an upper limit of temperature can easily be set and the changes or at any rate some of the changes responsible for damage or death can be clearly defined.

However, when the effect of cooling is examined it becomes very difficult to provide an exact definition of a lethal temperature. Certainly many forms of living cells can survive prolonged periods at very low temperatures. Luyet and Gehenio (1) have reviewed the effects of cold on plants, micro-organisms, protozoa, etc. When general cooling is considered, it is apparent that death takes place at temperatures considerably above the freezing point of protoplasm in all homeotherms and also in some poikilotherms. Cold exerts a continuing effect, so the duration of the cold exposure is of the utmost importance, owing to the disturbance of physiological functions, accumulation of toxic material, changes in viscosity, etc. In mammals, as pointed out in the section on hypothermia, death is essentially due to circulatory failure.

When local effects are concerned in man or other mammals, then the action of low temperatures on tissue is the predominant problem. The heart, except in hibernating animals, cannot survive temperatures below 10° to 18°C (50° to 64°F). But skin, muscle and nerve can certainly withstand short periods at temperatures that are in the range of 0° to 5°C (32° to 41°F) without any evidence of serious injury. As far as the living cell is concerned, it can survive extreme cold (−170°C (−274°F) or lower) under certain conditions. In the dry state micro-organisms can be kept for long periods at such temperatures, and on thawing will survive and multiply. Protozoa can also survive extreme cooling, if such cooling and subsequent thawing is very rapid.

Polge, Smith and Parkes (2) demonstrated that spermatozoa can be frozen to −79°C (−110°F) and rewarmed, and will still retain fertility. They have also demonstrated that portions of ovary can be successfully transplanted after freezing to −79°C (−110°F) for many days. Red cells suspended in glycerol have been kept at −79°C (−110°F) for up to 6 weeks and successfully transfused (3, 4).

It should be unnecessary to cite further evidence. It is clear that cold alone can be harmless to the living cell. However, it is also equally clear that low temperatures can be lethal, and the problem is why? The most obvious explanation is that the formation of ice crystals causes mechanical injury due to the expansion of ice. This injury will not be manifested except after thawing. Rapid thawing may diminish the injury: this has been explained by the type and rate of growth of crystals. With rapid freezing small crystals form from many foci. If freezing is slow, large crystals form, with greater mechanical effects. Slow thawing after rapid freezing leads to a phase of large crystal formation from small crystals, and hence damaging results. Finally the process of vitrification should be mentioned. Very rapid freezing prevents the formation of ice crystals, and the change of phase proceeds smoothly in either direction. Vitrification is only of academic interest in considering the effects of cold on man, as freezing time has to be measured in terms of milliseconds. The mechanical theory has been supported, amongst others, by Lewis (5), Greene (6), Lake (7), etc.

There is substantial evidence that the formation of ice crystals is not necessarily injurious. Red cells suspended in saline or plasma, haemolyse as a result of freezing and thawing. But when red cells are suspended in glycerol and cooled to −79°C (−110°F), crystallization of the medium certainly occurs, but there is no haemolysis on thawing. It is not certain, of course, that in these conditions freezing has taken place within the red cell. Luyet and Gehenio (1) quote other work to show that the mechanical injury story is difficult to support.

The most convincing theory of the damaging effect of crystallization is that which attributes ill effects to dehydration. As ice crystals form, so the remainder of the content of the cells will become more concentrated and dehydrated; and if this is prolonged irreversible changes occur. It can be tentatively concluded that the formation of ice crystals may be injurious, but that rapid or slow thawing and freezing do not affect the degree of injury. The injury might be mechanical, but there is no clear evidence for this hypothesis: the dehydration effect of crystallization is probably of considerable importance.

The threshold for cold injury may be proportional to the duration as well as the degree of cold. When the action of cold is brief, as in the use of the ethyl chloride spray for local anaesthesia, there is no evidence of injury. Crismon (8) measured the O_2 consumption of skin exposed at 2°C (35·6°F) for 90 minutes and found it was normal on returning to 37·5°C (99·5°F). But if skin was frozen at −2°C (28·4°F) the O_2 consumption did not return to normal on thawing. With moderately rapid freezing to −79°C (−110°F) and rapid thawing, there was a 50 per cent decrease in oxygen consumption; and it was found that the sodium content of the cells had increased five-fold. So it may be concluded that low temperature accompanied by crystallization can cause injury, but not necessarily death of tissue.

Trench-Foot and Immersion-Foot

However, when the large intact animal is considered, the problem has to be put another way. It is clear that serious injury can result from prolonged local cooling to temperatures which are well above freezing. It is necessary to emphasize that actual freezing is not necessarily locally lethal, and that on the other hand, much higher temperatures can be dangerous. What is the basis for moderate cold injury?

The 1914–18 war revealed the condition of trench-foot; in the 1939–45 war, immersion foot was described. They are certainly similar injuries and no advantage is to be gained from attempting to differentiate between the two. When persons are exposed to wet or damp cold conditions for many hours or days, damage to the extremities, particularly the feet, is likely to occur. Immersion foot was first observed and described carefully in shipwreck survivors, who had spent days on rafts with their feet and legs continually exposed to seawater, approximately at 8°C (46°F). Soldiers standing or sitting for many hours or days in wet trenches were liable to suffer from trench-foot. Ungley (9) has emphasized that the main factor is cold and that damage may occur with no more moisture than condensed sweat inside rubber boots. As nerve injury is a common feature, the alternative term 'peripheral vasoneuropathy after chilling' was proposed by Ungley. However, this phrase has not been generally adopted.

When patients are first examined, the feet are cold, swollen, and have a waxy appearance. There is difficulty in walking or balancing, the feet feel heavy and numb, and there are frequent descriptions such as 'it feels like walking on cotton wool'. This is the stage of ischaemia and is rapidly succeeded by the stage of hyperaemia, in which the feet are hot and red, with vigorous peripheral pulses, but no sweating. Swelling usually increases and blisters may form in the most severely damaged areas. Pain may be severe during this stage. Hyperaemia persists for many days or even weeks. Loss of sensation in the affected area is a common finding and the anaesthetic part is also anhidrotic. In severe cases gangrene develops peripherally. A full account of the various changes observed in immersion foot is given by Ungley (9). Apart from gangrene, the main damage is found in nerves and muscle. Complete degeneration of nerve trunks is not uncommon: muscle degeneration appears to be primary and not secondary to nerve injury. According to Ungley, vascular thrombosis is not an essential feature of immersion foot. There may be extensive damage to nerve and muscle without any evidence of vascular occlusion.

Friedman (10), on the other hand, who examined many cases describes considerable vascular changes. He emphasizes the great engorgement of blood vessels in the early stages, remarking that the vessels are as clearly defined as in tissues with vessels specially injected for teaching purposes. Medium sized vessels were dilated and with thin walls which sometimes ruptured. Red cells escaped from these vessels. Occasionally clumped masses of red blood cells were hyalinized and many vessels contained agglutinative thrombi. In material examined 40 days after injury, the vessels resembled those found in endarteritis obliterans. Degeneration of muscle was observed in early cases with extensive atrophy in late cases. Nerves obtained from the material in the inflammatory area were swollen, and oedematous, and nerve degeneration was frequently found central to the area affected. Friedman also comments on the frequency of fat necrosis and atrophy. The patients he examined had been exposed to temperatures of $0.5°$ to $4.5°C$ ($33°$ to $40°F$) with 96 per cent relative humidity. The men's feet became wet during landing operations on the Aleutians, and remained in damp boots for periods up to 6 days, but after landing were not immersed in water.

The main difference between Friedman and Ungley's observations is on the emphasis on vascular involvement. All workers have stressed the importance of the time factor.

Wet cold at temperatures considerably above freezing can produce severe and extensive damage in the extremities, if exposure is sufficiently prolonged. The pathological physiology is still somewhat obscure. When a foot or a hand is immersed in water at a temperature of 8°C (46·4°F) or lower, the blood flow is greatly increased (see Chapter 8). Experimentally such exposure has only been observed over a 3-hour period but vasodilatation persisted with periods of vasoconstriction throughout the 3 hours. However, the degree of cold vasodilatation is markedly affected by general body cooling, and the hand blood flow may not increase at all when the subject is cold, although in most cases there is a degree of cold vasodilatation even when the subject is severely chilled. Schwiegk (11) has emphasized the importance of general loss of body heat in the production of trench-foot or immersion foot. Ungley (9) quotes one case in which two men were in sea water at a temperature of −1·9°C (28·6°F), air temperature −20°C (−40°F) for 20 minutes, and then in an open boat for 10 minutes before rescue. One of the two subjects developed a moderately severe case of immersion foot. Local exposure without body chilling would not have produced any obvious damage.

The sequence of events would appear to be initial cold vasodilatation of varying degree in hands and feet. As body temperature falls and stored heat is lost, the cold vasodilatation is diminished and may be completely suppressed so that the blood flow is virtually nil. Ungley has remarked that the ischaemia must be partial or intermittent, as there is invariably considerable swelling of the foot before removal from the water. There must therefore have been some blood flow to permit sufficient exudation. Such swelling might well have taken place during the initial cold vasodilatation as there is considerable swelling even after 1 hour in water at 1° to 3°C (33·8° to 37·4°F), provided the part is dependent. Once cold vasodilatation is suppressed the temperature of the whole of the immersed part will cool to that of the surrounding water or very close to it. Precise limits of temperature that produce injury cannot be given with assurance. White (12) has described cases of survivors immersed in water as warm as 21°C (70°F) for many days, whose feet were swollen, hyperaemic and painful; it is not, however, certain that this condition was similar to immersion foot. Blood flow in limbs immersed in water at 20°C (68°F) is very small (13) and the temperature of the deep muscle, after 2 hours local immersion, may be only 1·5° to 2°C (2·7° to 3·6°F) above that of the water bath (Fig. 74). After 8 hours of such exposure the speed

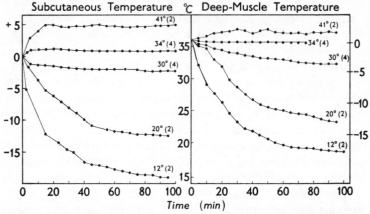

FIG. 74. The subcutaneous and deep muscle temperature in the forearm immersed in water at temperatures ranging from 12 to 41°C. Left hand and right hand ordinate gives the deviation from the initial temperature in the clothed forearm. The central ordinate gives the temperature in degrees centigrade. (From Barcroft and Edholm, *J. Physiol.*, **104**, 4, 366, 1946.)

of muscular movement is very greatly reduced, and this may be essentially due to the increased viscosity at lower temperature, or may be due to a direct affect on the neuro-muscular apparatus. The peculiar sensitivity of muscle to local chilling may be in part explained by the lack, or apparent lack, of cold vasodilatation in muscle vessels, which in turn may be due to the absence of *A.V.* shunts in muscle.

Conduction in mammalian nerve is markedly affected by temperature and conduction ceases at approximately 9°C (48°F). In hibernating mammals it is interesting that conduction continues in nerve fibres down to temperatures of 1° to 2°C (33·8° to 35·6°F). During the prolonged chilling required to produce immersion foot, blood flow will be greatly reduced and possibly there may be periods of complete ischaemia. The temperature of all tissues in the part exposed to cold water or wet cold will fall within 1 to 3 hours to a level close to that of the surrounding water, so nerve conduction may cease.

It would appear that the essential feature is a prolonged ischaemia at relatively low temperatures, i.e. temperatures from 1° to 15°C (34° to 41°F). Certainly mild cases can occur at the higher temperatures. In a recent trial for the Royal Navy, 9 volunteers were maintained in a covered rubber raft in Arctic waters for 5 days. Sea temperature averaged 0·5°C (32·9°F) and air temperature averaged 0°C (32°F). The subjects wore thick socks, boots, thick underclothes, seamen's trousers and jerseys. Air temperature inside the raft never fell below 10°C (50°F) and averaged 14°C (57°F). The floor of the raft was kept moderately

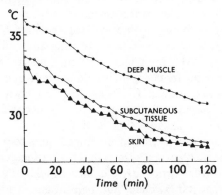

FIG. 75. The skin, subcutaneous and deep muscle temperature in the forearm exposed to room air (18°C (65°F)) at time 0. (From Barcroft and Edholm, *J. Physiol.*, **104**, 4366, 1946.)

dry by sponging: floor temperature was on occasions as low as 4°C (39°F). Toe temperatures were measured at intervals of 6 to 12 hours. The lowest recorded was 11°C (51·8°F) and most readings were of the order of 13° to 15°C (55·4° to 59°F). Although attention was paid to feet in that they were examined frequently and kept dry, only two subjects had completely normal feet at the end of the 5 days. The remainder had swollen, hyperaemic feet with some loss of sensation. In two subjects the hyperaemia and anaesthesia persisted for several weeks. These last two could certainly be described as mild cases of immersion foot.

The hyperaemia after exposure is reminiscent of the after dilatation described in detail by Wolff and Pochin (14) (see Chapter 8). When a finger is removed from an ice bath the vasodilatation may be due to metabolites, local damage with production of H substance, or vasoconstrictor paralysis. The increased blood flow is not affected by cold stimuli applied elsewhere. In immersion foot there is frequently some evidence of vasomotor paralysis after the hyperaemic phase has subsided. Anhydrosis which is usual in the hyperaemic area is probably due to degeneration of sudomotor fibres. After the hyperaemia has subsided the foot is frequently cold sensitive: blood flow is low and indirect vasodilatation with heat may be either slow or absent. Friedman's (10) findings indicate that the oedema fluid in immersion foot has a high fibrinogen content. This eventually clots and extensive fibrosis may therefore persist.

A milder form of cold injury is the chilblain. Lewis (5) considered that it is similar to trench-foot. It is not due to freezing, but to prolonged cooling with some body cooling. Chilblain subjects show normal vascular response to cold, and chilblains cannot be provoked easily by prolonged local cooling. It was found that in these subjects the foot or hand is cold not only in cold conditions but also in cool conditions. Chilblains are frequently seen in cases of peripheral vascular deficiency, such as occur in Raynaud's disease, syringomyelia and poliomyelitis (15).

Chilblains are usually found on fingers and toes and may be observed on ears. There appears to be a higher incidence in women than men, and it is particularly common in adolescents. The part affected becomes red, swollen, hot, painful and tender, and there is intense itching. In this hyperaemic stage there is considerable resemblance to mild immersion foot.

Frostbite

The most severe form of local injury is frostbite. Kreyberg (16) points out that very few cases of frostbite normally occur in cold countries, as the inhabitants understand how to live in a cold climate and how to avoid cold injury.

In war-time the position is very different. Very large numbers of men may be exposed to climatic extremes and cannot take the precautions that civilians normally would. The military significance of frostbite is very considerable. Although doubts have recently been cast on the generally accepted effect of the Russian winter on Napoleon in his retreat from Moscow, no doubts exist regarding the drastic effect of cold injury in the German forces in Russia in 1941–42. In the first World War, there were many thousands of cases of cold injury on the Western Front, apart from Russia. In the second World War, particularly during the Ardennes campaign in the winter of 1944–45, there were heavy casualties from cold in the British and Canadian forces.

In Korea, during the winter months, 25 per cent of all casualties in the American forces in the winter of 1950–51 were due to cold injury, mainly frostbite. So far as can be judged the casualties amongst the Chinese and North Korean troops were considerably greater, and in certain units there appeared to have been a 100 per cent incidence of cold injury.

Progress in knowledge of frostbite has been slow, owing to the paucity of cases in peace time, and the difficulty of careful study in war time. There have been no dramatic developments in treatment up to the present.

Kreyberg (16) summarized the state of knowledge at the outbreak of the second World War:

1. Local damage with necrosis of tissue is due to freezing of tissues or to prolonged exposure to temperatures just above 0°C (32°F).

2. There is a critical temperature below which tissue destruction is direct and immediate.

3. After thawing there are violent vascular reactions.

4. Vascular obstruction resulting in necrosis may be brought about by aggregation of red blood cells in the vessels concerned.

During exposure to very low temperatures, the intense vascular constriction virtually stops the circulation through the cooled part. If ice formation occurs, there is complete vascular occlusion. During thawing all vessels dilate maximally and the rate of blood flow is very high. In this stage of hyperaemia, the permeability of the vessels is increased, so there is considerable fluid loss from the blood. This stage has been studied in detail by Kreyberg using the technique of intravenous injection of fluorescein. The increased permeability may well be the cause of the increasing local haemo-concentration, which gradually leads to the blocking of vessels by the accumulation of red cells, so producing stasis with considerable oedema. The permeability of the vessels may be so great that red cells also escape into the tissue fluid. When stasis is complete, gangrene usually follows. However, Crismon (8) has observed that blood flow may persist in some thoroughfare channels, i.e. in arteriovenous shunts, for long periods after stasis is complete in other vessels. There may then be a gradual re-absorption of oedema fluid and a sufficient re-establishment of the circulation to prevent tissue destruction.

Q

This summary presents frostbite as essentially a vascular damage, the injury occurring during the stage of hyperaemia. Many authors subscribe to this view, including Lake (7), and therefore consider the object of treatment is logically to diminish the violence of the vascular reaction and particularly to control the permeability of the blood vessels.

However, other workers consider that cold damages or kills all or any tissue, and therefore the vascular change is only one of many. The treatment of frostbite which is adopted will depend on the view which is taken of the pathology. It may help to consider each stage in detail, citing such evidence as is available before considering treatment.

The Freezing of Tissues

In the first part of this chapter, evidence was given which showed that individual cells and certain tissues can survive actual freezing. The freezing point of skin, according to Lewis (5), is about $-1°$ to $-2°C$ (30° to 28°F). However, the phenomenon of supercooling is usually observed, when freezing does not occur until the surface temperature is lowered from $-5°$ to $-10°C$ (23° to 14°F). Occasionally even lower temperatures must be achieved, down to $-20°C$ ($-4°F$), before freezing takes place. Lewis stated that the fat content of the skin affected supercooling and if fat was applied to the surface then supercooling was more easily obtained. R. B. Lewis (17) measured the subcutaneous and muscle temperature in rabbits, when their legs were dipped into freezing mixtures. The temperature fell to approximately $-10°C$ (14°F) and then rose abruptly to approximately $-2.5°C$ (27.5°F) in the skin, which he considered was the true freezing point. Muscle froze at approximately the same temperature.

There is considerable controversy over the problem of whether actual freezing is itself lethal. Kreyberg (16) considers that low temperatures, whether accompanied by freezing or not, damage cells and tissues. In his view such damaged cells are then further injured during the phases of hyperaemia and stasis. But he does not consider that freezing is a lethal process comparable, say, to heat coagulation. Crismon (8) comes to similar conclusions in that he does not consider cold produces irreversible injury.

R. B. Lewis (17, 18) and his colleagues have produced experimental evidence to support the opposite view. They dipped the limbs of rabbits in freezing mixtures at $-12°C$ (10.4°F) for 30 minutes: 15 minutes after removal from the cold alcohol bath, the animals were sacrificed and microscopic signs of degeneration of the muscle cells were observed. These changes included actual disintegration of segments of muscle fibres. Connective tissue was more resistant than muscle fibres, but if observations were made on animals sacrificed at longer intervals after removal from the cold bath, then necrosis was observed in connective tissue as well. The changes in muscle included acute exudation in 4 to 6 hours, with the necrotic area well defined usually in 24 to 48 hours. On the other hand, vascular thrombosis was not observed before 24 hours and was not usual until 48 hours after the cold injury. Lewis therefore holds strongly to the view that cold injury directly damages or kills tissues, and that the vascular changes, including stasis, cannot be considered to be primarily responsible.

Crismon and his colleagues (8) have studied many different methods of treating experimental frostbite in animals. They find that immediate, rapid re-warming by dipping the frozen limb into hot water at a temperature of approximately 42°C (107·6°F), prevents gangrene, or greatly reduces loss of tissue. Such re-warming does not appear to influence the oedema of the frozen part. Fuhrman and Crismon (19) point out that the oedema fluid has a composition very similar to plasma, and that clotting may take place causing a fibrosis throughout the oedematous area. Rapid re-warming did not diminish damage due to fibrosis but there was diminished tissue loss. The effect of pressure dressings was tried, to control oedema with a view to preventing fibrosis. Their results were encouraging although others have not been able to confirm the value of pressure dressings.

Shumacker and his colleagues (20, 21) have also found that rapid re-warming diminished the incidence of tissue loss in experimental frostbite. The term 'rapid re-warming' is also used by the Russian workers, notably Aryev (22). However, the latter points out that really rapid re-warming of a frozen part is impossible owing to the low thermal conductance of human tissues. The Russians have used temperatures of the order of 35° to 36°C (95° to 96·8°F) for re-warming, and claim good results.

If the vascular reactions and subsequent oedema and stasis are mainly responsible for the destructive effect of frostbite, then methods of controlling the permeability of vessels should be tried. Fuhrman and Crismon (23) gave rutin to rabbits and obtained a diminished incidence of gangrene with a delayed onset. Stasis is delayed by rutin and fluid loss from the blood is slower. These effects do not appear to be due to vasoconstriction. Vasodilators, including CO_2, nitro-glycerine and mecholyl, were ineffective. Schwiegk (11) also concluded that damage occurs during re-warming. He therefore recommended that the chilled body should be warmed rapidly, but frostbitten extremities should be re-warmed slowly.

Laufman (24), on the other hand, is doubtful if the vascular changes are the sole agents responsible for death of tissue, at any rate in very severe frostbite. He found patent arteries, which spurted vigorously when cut in the midst of necrotic muscle, and as the necrotic muscle was covered by healing skin and fascia, he supports R. B. Lewis's views. Muscle is a much more specialized tissue than skin, with a higher resting metabolism, and is therefore more sensitive to low temperatures.

At present evidence is inadequate either to reconcile these two views, or to select one in favour of the other. It is necessary to point out some of the difficulties in accepting the experimental evidence presented so far without reservation. There is, first, the problem of species variation. What is true of the rat may not be true of the rabbit, and there are obvious problems in translating the results to man. Secondly, the methods employed included freezing with alcohol baths or CO_2 snow at very low temperatures. These experimental conditions are much more acute than those which are encountered naturally. The parts concerned also are small in relation to human limbs, e.g. the hind limbs of rats or rabbits, and in such experimental conditions the whole limb does indeed become frozen with ice crystals throughout the thickness of the limb. It may be noted at this point that Aryev (22) is emphatic that actual ice formation in frostbite in man does not occur. Aryev also denies that frozen parts are brittle, and he cites some experiments with frozen frogs thrown violently on to the ground, with no fractures. However, if a mouse is frozen at a sufficiently low temperature such as $-79°C$ ($-110°F$) it can be crumbled with ease (25).

Treatment of severe experimental local freezing of small laboratory animals with rapid re-warming in hot water significantly reduces the degree of tissue damage. Before such methods can confidently be applied to cases of human frostbite the differences to be expected need to be emphasized. In the experimental work, re-warming has been immediate, and since the parts concerned are small a fairly rapid rise of temperature can be expected throughout the frozen tissue, in spite of the poor thermal conductance of animal structures. Such rapid re-warming in man presents considerable practical difficulties, particularly in conditions of active service. More experimental work is required to determine the duration of the time interval after removal from freezing conditions in which rapid re-warming is still effective in reducing tissue loss. In man again, with feet and legs as the regions mainly affected, the rate of rise of temperature in the depth of the limb will be much slower in hot water than in the case of the small limbs of animals. For these reasons rapid re-warming in cases of frostbite in man has not yet been adopted.

The use of anti-coagulants in the treatment of frostbite has been advocated by several workers, notably Lange and his colleagues (26). There is some doubt as to the occurrence of true intravascular clotting in the stage of stasis, but agglutinative thrombi have been frequently reported. Lange has carried out a number of experiments both in animals and in man using heparin, in the expectation that there would be a significant diminution in vascular thrombosis. Significant reduction in the degree of tissue loss and in experimental frostbite in animals was reported by Lange, Weiner and Boyd (26). These workers also produced frostbite experimentally in man, and found reduction or absence of gangrene in cases treated with heparin.

These results could not be confirmed by other workers (20, 23, 27). Pichotka and Lewis (18) found that an alcohol bath at $-15°C$ (5°F) for 30 minutes was the highest temperature at which the majority of animals had partial or complete necrosis. Deep muscle temperature reached $-7°C$ (19·4°F), which Lake (7) considered to be the critical temperature for injury. Heparin had no effect on the degree of necrosis compared with control animals, but the death rate in the treated animals was significantly increased.

Although the weight of the experimental evidence is against the use of anti-coagulants, a recent report on clinical experience based on the treatment of 14 cases claims good effects (28). One of the points stressed in this report is that heparin must be started within 16 hours of the termination of cold injury. Since treatment as given included pressure bandages, oxygen therapy, lumbar block, etc. it is difficult to draw firm conclusions as to the part played by heparin.

Cortisone has also been used but the results obtained so far are not encouraging (24). Vasodilating procedures have been tried in order to improve the circulation through the affected area. Sympathectomy has been used by Shumacker (20 a, b) who considers that this method may have some value. The results are not clear cut, and more experimental work is needed to evaluate the rôle of sympathectomy.

Frostbite may have other than local effects. Aryev (22) considers that local cold injury always produces some general effects including myocardial changes. He also reports focal necrosis in liver, kidneys and suprarenals. Laufman found (24) in a case of extreme hypothermia with severe frostbite, clear evidence of renal damage with low sodium excretion and albuminuria. Water and electrolyte balance was severely disturbed, changes which were attributed to exhaustion of the suprarenal cortex. Laufman also reported that the E.C.G. showed disturbance of the T wave reminiscent of acute diffuse pericarditis.

This is a field which merits further experimental work, as treatment so far has been largely local.

Other Factors Complicating or Influencing Frostbite

The association of anoxia and frostbite has frequently been emphasized. During World War II, there were many cases of frostbite in high altitude flights. The incidence was considerably reduced when improved oxygen apparatus was installed. The conclusion was therefore drawn that anoxia increased the liability to frostbite. During acute anoxia, i.e. breathing 7 to 10 per cent oxygen at sea level, the peripheral blood flow is increased (29). Burton *et al.* (30) found in the rabbit an abrupt vasodilatation at an equivalent altitude of 17,000 ft. Such a vasodilatation was followed by a fall of rectal temperature. Anoxia, therefore, may affect the liability to frostbite by increasing heat loss and lowering body temperature, rather than by increasing the local effect of cold (31).

Recently, Pichotka, Lewis and Ulrich (32) have examined the influence of general hypoxia on local cold injury in rabbits. In the first group of experiments, the animals were exposed to a simulated altitude of 20,000 ft. for 30 minutes and then one hind limb was immersed in a freezing mixture of $-12°C$ $(10 \cdot 4°F)$ for 30 minutes. At the end of the total period of 60 minutes at high altitude, the animals were returned to ground level and the leg removed from the freezing mixture. No difference was observed between experimental and control animals in the degree of injury.

The effect of prolonged oxygen deficiency was then tested. Cold injury was started as soon as the animals reached a simulated altitude of 20,000 ft. The period of freezing was constant in all experiments, i.e. $-12°C$ ($10·4°F$) for 30 minutes. After removal of the hind limb from the freezing mixture the animals were maintained for 4 to 12 hours at altitude. With increasing time at altitude there was a significant increase in the degree of muscle injury. Controls showed 34 per cent muscle necrosis: after 4 to 6 hours hypoxia there was 48 per cent muscle necrosis which increased to 53 per cent in animals maintained 6 to 8 hours, and to 59 per cent when hypoxia was continued for 10 to 12 hours. In another experiment the duration of hypoxia preceding cold injury was varied from 0 to 5 hours, the animals being brought to sea level after the end of the freezing period. When the duration of hypoxia prior to injury was from 0 to 60 minutes, there was no change in the degree of necrosis. Sixty to 90 minutes prior hypoxia significantly increased the amount of damage, but if the hypoxic period was more prolonged the degree of cold injury declined again.

These interesting results emphasize the need for more work on the relationship between frostbite and anoxia. It is clear that continuing hypoxia after injury increases the degree of injury. Presumably, as Pichotka *et al.* suggest, tissues not irreversibly damaged by the cold injury are further damaged by the diminished oxygen supply. The duration of hypoxia before injury does not seem to be related to the extent of damage.

The factor in hypoxia which is more likely to cause frostbite is the mental deterioration and failure of judgment at high altitudes without oxygen. Individuals are liable to do the most foolish things. This point is illustrated by the experience of the French Expedition in the Himalayas recently. They successfully climbed a peak which is 27,000 ft. high, but they suffered very severe frostbite. On reaching the summit one of the climbers lost his gloves, which should of course have been tied to his wrist. Neither he, nor his companions, thought of using a spare pair of thick socks as a substitute, although that would have been normal procedure. On the way down, the party had to spend the night high up on the mountain. They removed their boots and left them outside their tent. In the morning the boots were lost in the snow which had fallen overnight. Much time had to be spent, in freezing weather, searching for the boots. As a result, the climbers suffered severe frostbite of both hands and feet with several amputations. Such carelessness in a party of skilled, brave and experienced mountaineers, can only be attributed to the known effects of oxygen lack.

Cold Injury of Respiratory Tract

Reports are not uncommon of pulmonary complications in subjects exposed to severe cold. Inhabitants of Alaska and the Canadian Northlands complain that their horses can suffer from 'frosting of the lungs'. However, careful search by competent medical observers has so far failed to obtain any definite evidence that local cold injury of the trachea or bronchi can occur. Armstrong and Burton (33) measured the temperature in the trachea just above the bifurcation, in the larynx and nose in dogs breathing room air and then very cold air. Even when the temperature of the inspired air was as low as −35°C (−31°F) little change was found in tracheal and laryngeal temperature. Arterial and venous pulmonary blood temperatures remained unchanged. It is evident that the upper respiratory tract has a very considerable rôle in temperature regulation and must have a very active vaso-motor system. Aschoff and others have shown that the temperature of the nasal mucosa changes more markedly than any other region in response to heating or cooling the surface of the body.

These experiments make it unlikely that even in very cold environments frostbite of the respiratory tract would occur. However, in conditions of heavy exercise with increased depth of ventilation, it is possible that there may be a fall of temperature within the respiratory tract. The sensitivity of asthmatics to breathing very cold air is well known. It is likely that the explanation lies in reflex effects mediated from the upper respiratory tract (34).

To summarize, severe damage can result from exposure of the extremities to moderate cold. There are many differences of opinion on the changes which occur, and it is impossible to be dogmatic about treatment. As in hypothermia, it is highly probable that very gradual or very rapid re-warming of the damaged part is safer than moderate re-warming. There is a great need for more detailed studies in this field which profitably could include more precise definition of the condition necessary to produce mild injury. A valuable discussion amplifying many of the points made in the chapter will be found in the Macy Foundation Conference on Cold Injury, 1952.

REFERENCES

1. LUYET, B. J. and GEHENIO, P. M. Mechanism of Injury and Death by Low Temperature. *Biodynamica*, **3**, 33, 1940–1941.
2. POLGE, C., SMITH, A. U. and PARKES, A. S. Revival of Spermatozoa after Vitrification and Dehydration at Low Temperatures. *Nature*, **164**, p. 666, 1949.
3. MOLLISON, P. L. and SLOVITER, H. A. Successful Transfusion of previously Frozen Human Red Cells. *Lancet*, **ii**, 862, 1951.
4. SMITH, A. U. Prevention of Haemolysis during Freezing and Thawing of Red Cells. *Lancet*, **ii**, 910, 1950.
5. LEWIS, T. Observations on some Normal and Injurious Effects of Cold upon the Skin and Underlying Tissues. *Brit. M. J.*, 795, 837, 869, 1941.
6. GREENE, R. Frostbite and Kindred Ills. *Lancet*, **ii**, 689, 1941.
7. LAKE, N. C. Effects of Cold. *Lancet*, **ii**, Oct. 13th, 1917.
8. CRISMON, J. M. Pathology and Physiology of Frostbite. *Bull. Vascular Surgery*, p. 110, 1951.
9. UNGLEY, C. C. The Immersion Foot Syndrome. *Advances in Surgery*, **i**, 269–336, 1949.
10. FRIEDMAN, N. B. The Pathology of Trench-Foot. *Am. J. Path.*, **21**, 387, 1945.
11. SCHWIEGK, H. Pathogenesis and Treatment of Local Cold Injury. 'German Aviation Medicine, World War II,' Surg. Gen. U.S.A.F., **2**, 843–857, 1950.
12. WHITE, J. C. Vascular and Neurologic Lesions in Survivors of Shipwreck. I. Immersion Foot Syndrome following Exposure to Cold. II. Painful Swollen Feet Secondary to Prolonged Dehydration and Malnutrition. In: Doherty, W. B. and Runes, D. C., eds. 'Rehabilitation of the War Injured', p. 647, Chapman and Hall, London, 1943.
13. BARCROFT, H. and EDHOLM, O. G. Temperature and Blood Flow in Human Forearm. *J. Physiol.*, **104**, 366, 1946.
14. WOLFF, H. H. and POCHIN, E. E. Vasodilatation After-Reaction in recently Cooled Fingers. *Clin. Sc.*, **8**, 145, 1949.
15. WINNER, A. and COOPER-WILLIS, E. S. Chilblains in Service Women. *Lancet*, **ii**, 663, 1946.
16. KREYBERG, L. Development of Acute Tissue Damage Due to Cold. *Physiol. Rev.*, **29**, 156, 1949.
17. LEWIS, R. B. Pathogenesis of Muscle Necrosis due to Experimental Local Cold Injury. *Am. J. M.Sc.*, **222**, 300-307, 1951.
18. PICHOTKA, J. and LEWIS, R. B. Muscle Necrosis after Experimental Freezing. U.S.A.F. School of Aviation Medicine, Report No. 7, 1951.
19. FUHRMAN, F. A. and CRISMON, J. M. Effect of Rutin in Experimental Frostbite in Rabbits. *J. Clin. Invest.*, **27**, 364, 1948.
20. SHUMACKER, H. B. (*a*) Sympathetic Interruption in Cases of Trauma. *Surg. Gynaec. & Obst.*, **84**, 739, 1947. (See also Wisconsin M.J.)
   (*b*) Sympathectomy in the Treatment of Frostbite. *Surg. Gynaec. & Obst.*, **93**, 727–734, 1951.
21. SHUMACKER, H. B. and LEMPKE, R. Recent Advances in Frostbite. *Bull. Vascular Surg.*, p. 77, 1951.

22. ARYEV, T. Ia. On the Question of the Pathology and Clinical Treatment of General and Local Hypothermia (a Résumé). *Klin. med. Mosk.*, **28**, (3) 15–24, 1950. Translation by E. R. Hope.
23. (a) FUHRMAN, F. A. and CRISMON, J. M. Studies on Gangrene Following Cold Injury. *J. Clin. Invest.*, **26**, 229, 236, 245, 476, 1947.
  (b) CRISMON, J. M. and FUHRMAN, F. A. Studies on Gangrene following Cold Injury. *J. Clin. Invest.*, **26**, 259, 268, 286, 1947.
24. LAUFMAN, H. Profound Accidental Hypothermia. *J.A.M.A.*, **147**, 1201, 1951.
25. PARKES, A. S. Personal Communication, 1952.
26. LANGE, K., WEINER, D. and BOYD, L. J. Frostbite. *New England J. Med.*, **237**, 383, 1947.
27. QUINTANELLA, R., KRUSEN, F. G. and ESSEX, H. E. Studies on Frostbite with Special Reference to Treatment and the Effect on Minute Blood Vessels. *Am. J. Physiol.*, **149**, 149, 1947.
28. THEIS, F. V., O'CONNOR, W. R. and WAHL, F. J. Anti-Coagulants in Acute Frostbite, *J.A.M.A.*, **146**, 992, 1951.
29. ANDERSON, D. P., ALLEN, W. J., BARCROFT, H., EDHOLM, O. G. and MANNING, G. W. Circulatory Changes during Fainting and Coma Caused by Oxygen Lack. *J. Physiol.*, **104**, 426, 1946.
30. BURTON, A. C., LAWSON, F. L., BROCK, L. and SMITH, S. The Effect of Anoxia on Vasomotor Tone and Temperature Regulation in the Rabbit. Associate Committee on Aviation Medical Research, Report No. C 2030, 1941.
31. LANGE, K., SCHÖTTLER, H. W. A., SCHÜTTE, E., SCHWIEGK, H. and WESTPHAL, V. Oxygen Lack in Frostbite. *Klin. Wchnschr.*, 444–653, 1943.
32. PICHOTKA, J., LEWIS, R. B. and ULRICH, H. H. The Effect of Hypoxia on Degree of Muscle Necrosis after Freezing. U.S.A.F. School of Aviation Medicine, Report No. 3, 1950.
33. ARMSTRONG, H. and BURTON, A. C. Unpublished observations, 1941.
34. DUKE, W. W. Physical Allergy, *J.A.M.A.*, **84**, 736, 1925.

CHAPTER 13

PROBLEMS FOR FUTURE RESEARCH

It has been suggested that it might be of interest to list what appear to be areas of ignorance in this field and pose questions for future research. Scattered throughout the text of the book, problems are mentioned, some of which are discussed below.

The material included in Chapter I describes some of the techniques of living in the cold. More work is required to enable the most suitable personnel to be selected for Arctic life: so far this has been done only on an ill-defined psychological basis. Clearly physiological adjustments are also of importance in order to live successfully in high latitudes. Are there any qualities or physique which particularly qualify an individual for this life? Do changes due to acclimatization outweigh such qualities? Small men or large, fat men or thin, which are the best? Certainly more work could be done to study the Eskimo and Indian in the North, the Tierra del Fuegans and the Sherpas of Nepal. There were very interesting observations made on the aborigines of Australia by Goldby, Hicks et al. (1938). In freezing weather, the aborigines spend the night lying virtually naked close to a fire. It was suggested that the heat given from the fire to the anterior surface of the body balanced the heat loss from the back. However more detailed experiments, using modern techniques, are clearly needed to work out the aborigines' mechanism of tolerance.

There are a considerable number of studies on the Eskimo and several have been mentioned in the text. These illustrate the great difficulty of generalizing about such peoples. The Eskimos are scattered over an enormous area, and each group lives with little or no contact with other groups. In consequence the cultural patterns differ widely and what is true of one group of Eskimos does not apply to another. This, in part at any rate, accounts for the very different opinions and statements regarding clothing, diet and shelter. These very differences provide opportunities for a greater variety of observations. Until recently the contact between Eskimos and other peoples was extremely small. Hence their physical characteristics have not been modified by inter-marriage and differences between groups could be studied and possibly related to diet, climate, etc. Just before this book went to press the Arctic Bibliography was published. This remarkable work contains 20,000 references dealing with all aspects of the Arctic, including the life of the Eskimo and other northern people.

There is considerable urgency if indigenous peoples are to be studied, as cultural contact with other civilizations is increasing rapidly. Such contacts can speedily modify the life of the previously isolated peoples—altering diet, clothing and shelter. The altered life may very well affect their cold tolerance.

The social aspects of such studies are important. There are vast areas of land in Northern Canada which formerly were virtually uninhabited. Recently, with the development of mineral resources, communities of considerable size are being established. The lack of knowledge of the long-term effects of cold on man implies that essential aspects of life in the north may have been neglected. It is difficult to obtain any detailed information concerning Europeans living permanently in the Arctic. There are problems which appear primarily concerned with social and economic factors, such as the length of service which is reasonable in the north; how frequent and how long should leave be; is it wise for children to be brought up entirely in the north; what are the risks and disadvantages of such life and hence what should the salary be. The proper answer to these and many other similar problems demands a knowledge of the physiological effects of cold climates.

There is one community or territory which has been studied in some detail, namely Alaska. In 1947, the population consisted of approximately 30,000 indigenous peoples and 60,000 whites. The medical difficulties are very considerable, and, as summarized by Barnett *et al.* (1947), include a high incidence of tuberculosis, venereal disease and alcoholism. Infant and child mortality appear to be very high. Many of the complaints made of life in Alaska, and indeed of the north in general, are similar to the grumbles usually heard in the tropics (Ellis, 1953): the loneliness and isolation, the enforced idleness imposed by the climate, the lack of amenities. It is clear that the physiological problems are not the only obstacles to extended settlement in the north. Nevertheless, a detailed investigation of first, second and third generation settlers in Alaska would be extremely valuable. Such a study should include anthropometric measurements; there may have been a change in body build according to the zoological rules of Berman and Allen, that is an increase in the girth of the limbs and a decreased linearity or shorter, fatter limbs. Although there have been many criticisms of these rules, a comparison of tropical people such as Africans with Eskimos appears to support them.

The effect of climate on the inhabitants of Siberia is discussed by Novakovsky (1942), who quotes vital statistics which may well by now be out of date. However many of his descriptions of the problems are similar to those in Alaska, particularly the enforced idleness of the long winter night.

The social patterns of life in the north deserve further research, including building questions which have been studied from engineering and military aspects. More needs to be known of the most desirable indoor temperatures that should be maintained. What advice can the physiologist give the builder on the lowest room temperature that can reasonably be tolerated? The answer will in turn be dependent on the insulation of the clothing worn. An approximate answer in functional terms could be that the individual must be warm enough to be able to sleep without shivering. Such an answer indicates the need for more precise information. Consideration of desirable or optimal standards as opposed to minimum standards also poses unanswered questions. Very high indoor temperatures, 27°C (80·6°F), are not uncommonly encountered in houses in the far north of Canada. A useful consequence is the considerable heat load in the clothing donned when going out. Such a heat load can delay heat loss from the body. There is no precise physiological basis for prescribing an indoor temperature of 27°C (80°F) or one of 15°C (60°F). There is a strong suggestion from Mackworth's work that cold acclimatization may be lost, or not acquired if the daily exposure to cold is too short, or if the temperature is too high during the rest of the day. So very warm houses may affect cold acclimatization. These problems are of considerable practical importance as they influence the cost of living in the Arctic, which can be very high.

It has been suggested above that more studies should be done on peoples living in the cold. Valuable, although limited information has come from members of Arctic expeditions. It is worth emphasizing that useful physiological data can only be obtained by professional physiologists, as the observations required on Arctic expeditions demand skill and knowledge. It has unfortunately been assumed only too frequently that physiological information can easily be provided by casual measurements.

Amongst the problems that can usefully be studied in the field, as well as in the laboratory, are those of clothing. Laboratory assessment does not always coincide with field assessment. This may in part be due to prejudice, but probably also indicates the need for improved laboratory technique.

The value for the insulation of air when the body is in motion is significant, and the changes in the insulation of clothing during movement have not yet been systematically examined. The features in clothing design which encourage a decreased insulation during exercise need further exploration.

Factors which determine the thermal insulation of a given substance or tissue are still to a large extent unknown, and the field appears to be neglected by contemporary physicists. The thermal conductivity of biological material such as muscle is highly variable even when removed from the body, and the variability does not seem to be related to the fat or water content. The effect of fat upon the skin deserves further study; it seems unlikely that the very thin layer of grease which long distance swimmers apply to their bodies can materially affect their heat loss, but great subjective relief from cold is claimed. In such studies, and indeed in many others, the heat balance of the body has to be assessed, and this can only be satisfactorily treated in the steady state.

When body temperatures are changing we are very far from being able to calculate accurately the heat exchange, in spite of using the device of skin temperature and core temperature. Still less are we able to make such calculations when clothing is worn. Much work is required using simultaneous direct and indirect calorimetry. Physiologists and biochemists assume that heat production can be calculated from oxygen consumption and the respiratory quotient: the assumption appears to be soundly based on the classic work of Rubner, Voit, Lusk, Cathcart, etc. However, when metabolic changes other than the complete oxidation of protein, carbohydrate or fat occur, the calories/litre of oxygen are unknown. In hibernation the R.Q. appears to remain for long periods at values of $0 \cdot 4$; what is the calorie value per litre of oxygen under such conditions?

The problem of the heat balance of Arctic mammals has been clarified by the brilliant work of Irving and Scholander and the outstanding difficulty now is to determine how such animals increase their heat loss when the temperature moderates. Is this achieved primarily by a vascular re-adjustment? Amongst vascular reactions to cold, cold vasodilatation has been much studied but the mechanism remains unknown. Are there any modifications in this response in those adapted to live in the cold, e.g. in the Eskimo? The rôle of the venae comites needs further study, especially the possible shift in venous return under different thermal conditions.

A problem in the metabolic field is to inquire into the reasons for the increased food intake in the cold. There are many contradictory reports concerning the amount of food consumed in the north, and some recent and very detailed studies by Rodahl cast doubts on the validity of observations of high calorie intakes. Studies in Greenland by Høygaard on Eskimo families for many months signify dietary intakes similar or smaller as regards calories to those in temperate climates. It is important to stress the variations in metabolic expenditure and calorie intake in the many possible situations in the north. Detailed continuous studies of energy expenditure and calorie intake are required to settle this important practical point.

Another metabolic problem which deserves closer study is that of shivering. It is not clear whether there is a definite march of activity over the muscles of the body, or whether the muscle groups involved are determined by local chillings. The variation in intensity has not yet been satisfactorily explained, nor have the factors influencing the rate of fatigue of shivering. As the body cools, after an initial period of violent shivering, there may succeed a sustained muscular rigidity, of which little is known.

It will be evident from Chapter 10 that there are many unsolved problems in the field of acclimatization. Much work has been and is being done on the rat, some on the rabbit and a little on man: results are not necessarily interchangeable. Many valuable experiments can be conducted without using expensive and elaborate cold chambers. Few investigations have as yet been done on those habitually exposed to the cold, such as outdoor workers, including farm workers, truck drivers, fishermen, seamen, etc. To obtain cold stress it is not necessary to have subjects dressed in Arctic clothing at −40°C.

There are a number of specific physiological problems mentioned in the text, including those revealed by investigations on hypothermia. The latter have now assumed a more urgent aspect, as the clinical use of hypothermia is increasing.

The surgical treatment of congenital cardiac malformations is one of extreme difficulty and the mortality is considerable. It is usually necessary to clamp one or more of the main vessels leaving or entering the heart for a considerable period. The possibility of doing this successfully is greatly increased by lowering body temperature to approximately 26°C (78°F), when oxygen consumption is only 20–25 per cent of the normal resting value and the heart rate may be 15–20 beats per minute. A considerable number of operations have already been done on hypothermic patients, and there is little doubt that some cases have been treated satisfactorily who could not have survived operation under more ordinary conditions. It seems also that a number of patients have died who might have lived if treated conventionally. Hypothermia is still a dangerous state and so is the re-warming phase. Ventricular fibrillation is the most serious complication, although defibrillation can be successfully accomplished by electrical stimulation. Further work to establish the cause of cardiac irregularity during hypothermia is urgently needed.

There seems to be convincing evidence that rapid re-warming is the preferred treatment of hypothermia, but we still lack proof of the value of this treatment when hypothermia is prolonged and there may have been considerable shifts of body water. Circulatory collapse during re-warming is also a complication. A brief description of such failure is given in Chapter 11, based on observations made on cases of prolonged therepautic hypothermia. It is not certain if the cases which are now described are due to a sudden peripheral vasodilatation, and further experimental work is needed. If this potentially valuable method of treatment is not to be condemned, it is important to carry out more studies and for the clinicians to exercise restraint. Various drugs, such as largactol, are employed to increase or facilitate the drop in body temperature, but the mode of action is far from clear.

The biochemical changes in hypothermia have so far been neglected, and this should prove a fruitful field of work as indeed should the biochemical changes in all aspects of low temperature. The mode of action of acclimatization certainly requires to be studied at the cellular level. There are already some studies to show modification of enzyme systems in the liver in cold acclimatized rats, and it seems probable that there are many other examples remaining to be discovered. In true hibernation the biochemical aspects have not yet been thoroughly investigated.

The problems of cold injury, in spite of a vast literature, are virtually unsolved. The mildest forms of injury such as the chilblain have not yet received their proper share of attention. These seemingly trivial conditions deserve more research as their investigation may well give valuable clues regarding more serious states, such as frostbite. There is a wealth of clinical observation on frostbite but it is largely uncontrolled. Fundamental problems such as the mechanism of crystal formation, the interplay of mechanical and vascular reaction still await elucidation.

We may summarize the effects of low temperature on man throughout the world by repeating a famous witticism. 'Many are cold but few are frozen.'

REFERENCES

1. GOLDBY, F., HICKS, C. S., O'CONNOR, W. J. and SINCLAIR, D. A. A Comparison of the Skin Temperature and Skin Circulation of Naked Whites and Australian Aborigines Exposed to Similar Environmental Changes. *Australian J. Exper. Biol. & M.Sc.*, **16**, 29–37, 1938.
2. Arctic Bibliography, Department of Defence, U.S.A. 3 vols. 1953.
3. BARNETT, H. E., FIELDS, J., MILLES, G., SILVERSTEIN, J. and BERNSTEIN, A. Medical Conditions in Alaska. *J.A.M.A.*, **135**, 500, 1947.
4. NOVAKOVSKY, S. (a) The Probable Effect of the Climate of the Russian Far East on Human Life and Activity. *Ecology*, **3**, 181, 1922.
   (b) The Effect of Climate on the Efficiency of the People of the Russian Far East. *Ecology*, **3**, 275, 1922.
5. ELLIS, F. P. Tropical Fatigue. Symposium on Fatigue. Edited by W. F. Floyd and A. T. Welford. London, H. K. Lewis & Co., Ltd., 1953.

AUTHOR INDEX

A

ADOLPH, E. F.
'Absence of acclimatization to cold in man,' 187
'Hypothermia in infant animals', 7
'Hypothermia in acclimatized rats', 175
'Increased cold tolerance in new born animals', 175
'Posture and cold diuresis', 143
'Rectal temperature and heart rate', 207

ALEXANDER, L.
'Hypothermia experiments at Dachau', 205

ALLEN, W. J.
'Anoxia on peripheral blood flow', 236

ALPER, J. M.
'Inhibition of shivering by inhaling CO_2', 155

AMES, A.
'Increased tolerance to cold in man', 182

ANDERSON, D. P.
'Effect of anoxia on peripheral blood flow', 236

ANDJUS, R.
'Resuscitation of rats cooled to 1°C.', 216

ARCTIC BIBLIOGRAPHY, 242

ARMSTRONG, H.
'Evaporation from upper respiratory tract', 30
'Inhalation of cold air on respiratory tract temperature', 238

ARRHENIUS, Law of, 3

ARYEV, T. IA.
'Crystal formation not found in frostbite', 234
'General effects of local cold injury', 236
'The effect of rapid rewarming in frostbite', 233

ASCHOFF, J.
'Effect of body cooling on nasal mucosa', 238
'Gradient of blood flow in extremities', 133
'Vasodilatation in cold fingers', 131

ASTWOOD, E. B.
'Thyroid secretion in the cold', 166

B

BADER, M. E.
'Body temperature and hand circulation', 139
'Effect of face heating on hand blood flow', 140

BADER, R. A.
'Negative chloride balance during cold diuresis', 143
'Posterior pituitary and cold diuresis', 144
'Reactions to cold in heat acclimatized subjects', 183

BALKE, B.
'Peripheral blood flow in cold acclimatized subjects', 188

BARCROFT, H.
'Blood flow in cooled limbs', 228
'Effect of anoxia on peripheral blood flow', 236
'Effect of local temperature on vascular responses', 140
'Temperature and forearm blood flow', 129

BARNETT, H. E.
'Medical conditions in Alaska', 242

BASS, D. E.
'Effect of acclimatization to heat on reaction in the cold', 183
'Posterior pituitary and cold diuresis', 144

S

s*

SUBJECT INDEX

A

Aborigine, acclimatization to cold in Australian, 241
Acclimatization, definition of, 162
Acclimatization to cold (see Cold Exposure)
  in Antarctic expeditions, 189
  and ascorbic acid, 168
  ascorbic acid, excretion of, 168
  in Australian aborigines, 241
  in clipped rats, 173
  cold chamber experiments, 183
  cutaneous dehydration, 189
  and heat, 243
  and high fat diet, 181
  hypertension and ascorbic acid, 169
  hypertension in, 167
  liver O_2 consumption, 172
  local, in hands, 190
  in man, 187
  in man, mechanism of, 188
  metabolism and, 169
  metabolism, mechanism of increased, 171
  metabolism in rats, 173
  muscular exercise and, 176
  and nutrition, 180
  and peripheral circulation, 188
  in rabbits, 174
  and severe exposure, 186
  suprarenals, 166
  in swimmers, 212
  thyroid secretion, 166
  and thyroidectomy, 172
  in Tierra del Fuegans, 15
  vitamin intake and, 181
  weight changes in, 167
Acclimatization to heat and response to cold, 183
Acclimatization to temperature, in fish, 163
Acetyl-choline in hypothermic brain, 205
Adaptation to cold, in Arctic animals, 94, 164
  increased insulation, 96
Adrenaline, calorigenic action of, 149
  secretion in cold, 148
Air
  thermal insulation, 28, 36, 47, 49, 56
  thermal insulation and altitude, 52
  thermal insulation, changes in, 99
  thermal insulation and wind, 51
Air Movement (see also Wind)
  and clothing insulation, 64
  and heat loss, 49
Alarm reaction, 167

Alaska, medical conditions, 242
Altitude and insulation of air, 52
Anaesthesia in hypothermia, 208
  and shivering, 155
Anoxia and frostbite, 236
Anoxia
  and peripheral circulation, 236
  and shivering, 154
Antarctic, cold tolerance in, 189
Arctic animals, adaptation to cold, 164
  adaptation by insulation, 94
  critical temperature, 93
  temperature regulation, 91
  thermal insulation, 63
Arrhenius, Law of, 3
Arterial blood, temperature changes, 142
Arteriovenous anastomoses
  cold vasodilatation in fingers, 131
  in fingers, 133
Ascorbic acid, and acclimatization to cold, 168
  excretion on exposure to cold, 168
  and hypertension in cold, 169
  and resistance to cold, 169
  in suprarenals, 168
  in tissues after cold exposure, 167
  and utilization of cortical hormones, 169
Asthma, inhalation of cold air, 238
Atropine and hypothermic bradycardia, 208
Australian aborigine, effect of cold on, 241
Axon reflex in cold vasodilatation, 131

B

Basal metabolism (see Metabolism)
  in the Arctic, 178
Bat, body temperature, 92
  temperature, regulation of, 10
Bath calorimeter, 84
Bath temperature and metabolism, 154
Blood, low temperature storage, 224
Blood flow (see Circulation)
  in cooled limbs, 228
  of fingers, 73
  fingers in cold, 131
  and frostbite, 231
  gradient in extremities, 133
  in hands of Eskimos, 192
  in hand, on heating face, 140
  liver, in hypothermia, 215
  temperature and, 129
  vasodilatation in cold air, 135
Blood gases in hypothermia, 213
Blood pressure (see Hypertension)

Blood pH in hypothermia, 214
Blood sugar in hypothermia, 214
Blood temperature
  arterial and venous temperatures, 86
  in right and left auricle, 201
Blood volume, and climate, 184
  and cold diuresis, 143
Blood volume
  and cold exposure, 187
  and hypothermia, 204, 218
Body cooling, effect on man, 209
  and subcutaneous fat, 212
  vasoconstriction in nasal mucosa, 143
Body temperature
  of bats, 92
  formula for calculating, 39
  and hand blood flow, 133
  hand circulation, 139
  and heat balance, 12
  lethal limits, 13
  lethal limit in animals, 206
  lethal limits in man, 208
  maintenance in cold, 91
  measurement of, 200
  need for constancy, 6
  physical regulation, 8, 148
  range in animals, 11
  of sloth, 92
  stability of regulation, 13
  variations in, 14
  variations in core, 200
  volume of superficial tissues, 14
  why regulated at 37°C., 11
Body tissues, thermal insulation of whole body, 84
Body warming and hand circulation, 139
Boots, vapour permeability of, 70
Buildings, humidity and insulation of, 67

C

Calorie intake (see also Nutrition)
  of Eskimo, 179
  increase at low temperatures, 165, 185
  related to temperature, 157
Capillary filtration, 69
Carbon dioxide content of blood in hypothermia, 213
Cardiac output (see Heart)
  climatic effects on, 185
Channel swimmers and cold tolerance, 212
Chemical heat regulation, 148
Chemical reactions and temperature, 3
Chemical regulation, evidence for, 172
  in the cold, 156, 175
Chilblains, description, 230

Circulation (see also Peripheral circulation and blood flow)
  axon reflex in cold vasodilatation, 131
  and body warming, 139
  climatic effects on, 185
  cold exposure and blood volume, 187
  cold pressor test, 138
  peripheral resistance in hypothermia, 218
  thermal circulation index, 81
  vasoconstriction in nasal mucosa in cold, 143
  vasodilatation after exposure to cold, 137
  vasodilatation in fingers in the cold, 133
  venous drainage from cold vasodilated hands, 142
Civilization, climate and, 18
  coldward movement of, 18
  and homeothermy, 16
Climate, civilization and, 18
  effects in Siberia, 243
  energy of nations, 16
  thermal analysis of, 118
Climatic stress, social problems, 242
Clo unit, definition of, 35
Clothing (see Textiles)
  air permeability of, 64
  curvature on insulation of, 60
  design of, 59
  frost line in, 69
  greenhouse effect, 125
  heat distribution in, 140
  heat flow through, 28
  and heat loss by evaporation, 71
  heat regain, 66
  insulation of assembly, 63
  maximum insulation, 63
  metabolic cost of Arctic, 99, 158
  moisture and heat transfer, 66
  reflecting power, 122
  relative humidity in, 68
  vapour resistance, 37, 70
  walking and insulation of, 64
  and wind chill, 111
  wind and insulation of, 64
Cold diuresis, blood volume changes, 143
  negative chloride balance in, 143, 185
  posterior pituitary and, 144
  posture and, 143
Cold exposure (see Acclimatization to cold)
  adrenaline secretion, 148
  and blood volume, 185
  and haemoconcentration, 187
  nutritional factors reviewed, 181